高等学校应用型本科创新人才培养计划指定教材

高等学校计算机类专业"十三五"课改规划教材

Android 高级开发及实践

青岛农业大学
青岛英谷教育科技股份有限公司 编著

西安电子科技大学出版社

内 容 简 介

本书在“Android 程序设计及实践”课程的基础上，以理论联系实践的形式深入地讲解了 Android 高级开发的相关知识与技术。全书共有 8 章，具体介绍了 Content Provider、图形图像与动画、高级网络编程、高级用户体验、传感器、Wi-Fi 与 Bluetooth、NFC 以及资源与国际化等知识。另外，本书还讲解了移动物联网的相关概念及有关程序的实现、NFC 近场通信技术等。

本书案例基于 Eclipse 开发工具编写，使用的 SDK 版本为 Android4.3(API 18)。

本书适用范围较广，可作为高等院校计算机科学与技术、移动互联网、软件工程、网络工程、计算机软件、计算机信息管理以及电子商务等专业的程序设计课程的教材，也可作为科研、程序设计等人员的参考书籍。

图书在版编目(CIP)数据

Android 高级开发及实践/青岛农业大学，青岛英谷教育科技股份有限公司编著.
—西安：西安电子科技大学出版社，2016.11
高等学校计算机类专业“十三五”课改规划教材
ISBN 978-7-5606-4316-8

Ⅰ. ① A… Ⅱ. ① 青… ② 青… Ⅲ. ① 移动终端—应用程序—程序设计—高等学校—教材
Ⅳ. ① TN929.53

中国版本图书馆 CIP 数据核字(2016)第 269047 号

策　　划　毛红兵
责任编辑　万晶晶　毛红兵
出版发行　西安电子科技大学出版社(西安市太白南路 2 号)
电　　话　(029)88242885　88201467　　邮　　编　710071
网　　址　www.xduph.com　　电子邮箱　xdupfxb001@163.com
经　　销　新华书店
印刷单位　陕西天意印务有限责任公司
版　　次　2016 年 11 月第 1 版　2016 年 11 月第 1 次印刷
开　　本　787 毫米×1092 毫米　1/16　印　张　19
字　　数　447 千字
印　　数　1～3000 册
定　　价　46.00 元

ISBN 978-7-5606-4316-8/TN

XDUP 4608001-1

❖❖❖ 前　言 ❖❖❖

本科教育是我国高等教育的基础，而应用型本科教育是高等教育由精英教育向大众化教育转变的必然产物，是社会经济发展的要求，也是今后我国高等教育规模扩张的重点。应用型创新人才培养的重点在于训练学生将所学理论知识应用于解决实际问题，这主要依靠课程的优化设计以及教学内容和方法的更新。

另外，随着我国信息技术的迅猛发展，社会对具备信息技术能力的人才需求急剧增加，“全面贴近企业需求，无缝打造专业实用人才”是目前高校计算机专业教育的革新方向。为了适应高等教育体制改革的新形势，积极探索适应 21 世纪人才培养的教学模式，我们组织编写了高等院校计算机专业系列课改教材。

该系列教材面向高校计算机专业应用型本科人才的培养，强调产学研结合，经过了充分的调研和论证，并参照多所高校一线专家的意见，具有系统性、实用性等特点。旨在使读者在系统掌握计算机知识的同时，着重培养其综合应用能力和解决问题的能力。

该系列教材具有如下几个特色：

1. 以培养应用型人才为目标

本系列教材以应用型软件人才为培养目标，在原有体制教育的基础上对课程进行了改革，强化“应用型”技术的学习。使读者在经过系统、完整的学习后能够掌握如下技能：

- ✧ 掌握信息系统开发所需的理论和技术体系以及系统开发过程规范体系；
- ✧ 能够熟练地进行信息系统设计和编码工作，并具备良好的自学能力；
- ✧ 具备一定的项目经验，包括代码的调试、文档编写、软件测试等内容；
- ✧ 达到信息技术企业的用人标准，做到学校学习与企业的无缝对接。

2. 以新颖的教材架构来引导学习

本系列教材采用的教材架构打破了传统的以知识为标准编写教材的方法，引导读者在学习理论知识的同时，加强实践动手能力的训练。

教材内容的选取遵循“二八原则”，即重点内容由企业中常用的 20%的技术组成。每个章节设有本章目标，明确本章学习重点和难点，章节内容结合示例代码，引导读者循序渐进地理解和掌握这些知识和技能，培养学生的逻辑思维能力，掌握信息系统开发的必备知识和技巧。

另外，本系列教材借鉴了软件开发中的“低耦合，高内聚”的设计理念，组织结构上遵循软件开发中的 MVC 理念，即在保证最小教学集的前提下可以根据自身的实际情况对整个课程体系进行横向或纵向裁剪。

3. 提供全面的教辅产品来辅助教学实施

为充分体现“实境耦合”的教学模式，方便教学实施，该系列教材配备可配套使用的项目实训教材和全套教辅产品。

- ✧ 实训教材：集多线于一面，以辅助教材的形式，提供适应当前课程(及先行课程)的综合项目，遵循系统开发过程，进行讲解、分析、设计、指导，注重工作过程的系统性，培养读者解决实际问题的能力，是实施“实境”教学的关键环节。
- ✧ 立体配套：为适应教学模式和教学方法的改革，本系列教材提供完备的教辅产品，主要包括教学指导、实验指导、电子课件、习题集、实践案例等内容，并配以相应的网络教学资源。教学实施方面，提供全方位的解决方案(课程体系解决方案、实训解决方案、教师培训解决方案和就业指导解决方案等)，以适应信息系统开发教学过程的特殊性。

本书由青岛农业大学和青岛英谷教育科技股份有限公司共同编写，参与本书编写工作的有吕健波、邵作伟、宁维巍、宋国强、侯方超、何莉娟、王千、杨敬熹、刘江林、王万琦等。本书在编写期间得到了各合作院校专家及一线教师的大力支持与协作，在此，衷心感谢每一位老师与同事为本书出版所付出的努力。

由于编者水平有限，书中难免有不足之处，欢迎大家批评指正！读者在阅读过程中发现问题，可以通过邮箱(yinggu@121ugrow.com)发给我们，以期进一步完善。

本书编委会

2016 年 9 月

❖❖❖ 目　　录 ❖❖❖

第1章 Content Provider

本章目标

- 了解 Content Provider 运行机制
- 理解 Content Provider 共享数据的规则
- 掌握获取系统通讯录的流程
- 掌握自定义 Content Provider

Content Provider(内容提供者)是 Android 系统的四大组件之一，使用 Content Provider 组件可以实现多个程序之间数据的存储和读取。本章主要讲解如何通过 Content Provider 获取系统通讯录，以及如何在自己的程序中自定义 Content Provider 组件，以向外界提供数据。

1.1 Content Provider 概述

虽然 Android 为每个程序开辟了一块独立的内存区域，却没有提供一个公共的内存区域以供程序之间共享数据。Content Provider 组件恰好可以解决这一问题。

通过 Content Provider 组件能够实现如下功能：

- ✧ 访问系统程序或其他程序提供的数据，例如系统通讯录、短信等；
- ✧ 开发的程序中，有部分数据可供给其他程序操作。

如图 1-1 所示，程序 A 提供了 Content Provider 接口，这个接口允许其他程序对程序 A 中的数据进行操作，这些数据可以是普通文件、XML 文件、数据库或网络；程序 B 只需要按照一定的规则访问 Content Provider 接口，即可获取相应的数据。在这里，程序 A 只需公开可共享的数据，程序 B 通过接口即可获取自己关心的数据。

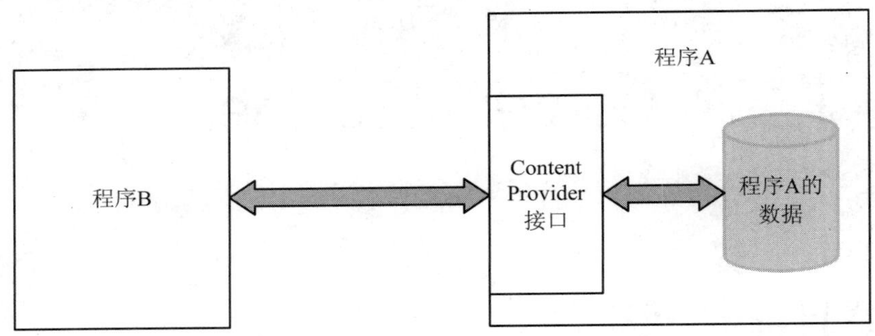

图 1-1 Content Provider 实现流程图

1.1.1 相关 API

Android API 提供了一系列的类来实现或操作 Content Provider 相关功能，主要涉及以下几类：

1. Content Provider

Content Provider 是内容提供程序的基类。该类负责将要共享的数据进行封装，向外界程序提供统一的数据接口，实现数据共享。数据接口使用 Uri 协议规范，通常实现一个内容提供程序，需要定义多个 Uri 接口。需要注意的是：Content Provider 类只能作“内容提供者”，而非“使用者”。

创建一个自定义的内容提供程序，需要继承 Content Provider 类，重写该类的方法如下：

- ✧ onCreate()：只执行一次，通常用于初始化必要的程序设置，例如创建数据库实例等。
- ✧ insert(Uri uri, ContentValues values)：插入数据。把新数据插入内容提供程序中，返回 Uri 对象，表示新插入项的 URI。参数如下：
  - ➢ uri：操作数据的 URI。
  - ➢ values：表示一条数据的参数，以“键值对”格式存放。
- ✧ delete(Uri uri, String selection, String[] selectionArgs)：删除数据。删除内容提供程序中的数据，返回 int 类型值，表示删除操作受影响的行数。参数如下：
  - ➢ selection：对应 SQL 语句中的条件部分，表示删除数据的条件。
  - ➢ selectionArgs：条件语句中对应的占位符填充。
- ✧ update(Uri uri, ContentValues values, String selection, String[] selectionArgs)：更新数据。更新内容提供程序中已有的数据，返回 int 类型值，表示更新操作受影响的行数。
- ✧ query(Uri uri, String[] projection, String selection, String[] selectionArgs, String sortOrder)：查询数据。返回相应的查询结果，返回 Cursor 游标对象。参数如下：
  - ➢ projection：指定要查询的列，如果为空，则表示查询所有列。
  - ➢ sortOrder：指定查询的排序规则。
- ✧ getType(Uri uri)：查询 URI 类型。返回 String 类型值，是传入 Uri 对象的 MIME 数据类型。此数据类型以“vnd.android.cursor.item/”开头，表示一条记录；以“vnd.android.cursor.dir/”开头，表示多条记录。

2．Content Resolver

Content Resolver 类在 Content Provider 中担任“使用者”的角色，用于根据指定的 URI 操作对应的内容提供程序，实现数据的“增删改查”操作。因此，Content Resolver 的主要方法如下：

- ✧ insert(Uri url, ContentValues values)；
- ✧ delete(Uri url, String where, String[] selectionArgs)；
- ✧ update(Uri uri, ContentValues values, String where, String[] selectionArgs)；
- ✧ query(Uri uri, String[] projection, String selection, String[] selectionArgs, String sortOrder)。

以上四个“增删改查”方法，与 Content Provider 中对应方法的功能是类似的，区别在于：Content Provider 是对内容提供程序本身的数据进行操作的，而 Content Resolver 用于操作其他程序提供的 Content Provider。

3．ContentValues

ContentValues 类用于以“键值对”的格式存储一系列数值，通常描述数据表中一条记录。该类常用方法如表 1-1 所示。

表 1-1　ContentValues 类常用方法

方法名	描述
put(String key, String value)	添加一个 String 类型的数值，该方法为重载方法，支持一系列的基本数据类型数值的添加
putAll(ContentValues other)	添加 other 中所有值
putNull(String key)	添加一个空值
Object get(String key)	获取一个值
String getAsString(String key)	获取一个 String 类型值，与 put()方法相对应。与之类似的还有 getAsInteger()、getAsBoolean()、getAsByteArray()等方法
remove(String key)	删除一个值
clear()	清除所有值
size()	获取值的数量
Set<String> keySet()	获取“key”的集合
boolean containsKey(String key)	查看是否存在与“key”对应的值

4．ContentUris

ContentUris 类用于处理 Uri 对象，针对 URI 提供了以下常用方法：

- ✧ Uri withAppendedId(Uri contentUri, long id)：静态方法，为 Uri 对象附加给定的 ID，返回 Uri 对象。
- ✧ long parseId(Uri contentUri)：静态方法，解析给定的 Uri 对象，返回解析后的 long 类型 ID 值。

5．UriMatcher

UriMatcher 类是一个工具类，用于匹配 Content Provider 中用到的 URI。常用方法如下：

- ✧ UriMatcher(int code)：构造方法，“code”为 URI 的根节点，通常传入常量“UriMatcher.NO_MATCH”。
- ✧ addURI(String authority, String path, int code)：添加一个 URI 匹配规则。参数如下：
  - ➢ authority：权限匹配部分。
  - ➢ path：路径，通常表示要操作的数据表的名称。
  - ➢ code：匹配成功的返回码。
- ✧ match(Uri uri)：尝试匹配 Uri 路径，返回 int 类型数值，表示匹配的“code”。

1.1.2　Content Provider 操作规则

访问程序提供的 Content Provider，必须遵循一个规则，就像浏览网页时必须遵循 HTTP 协议一样。

Content Provider 通过 Uri 对象来实现数据共享，相当于浏览网页时输入的域名。标准 URI 如下所示：

```
content://authority/path/id
```

- ✧ 第一部分(content)：协议头，必须以“content://”开头，对应的常量为“ContentResolver.SCHEME_CONTENT”。
- ✧ 第二部分(authority)：权限部分，要保证每个 Content Provider 具有唯一的“authority”，权限部分必须与 AndroidManifest.xml 文件中声明 ContentProvider 时的“authority”保持一致。
- ✧ 第三部分(path)：资源部分，多个层次使用“/”分隔，通常用于表示要操作的数据表的名称。
- ✧ 第四部分(id)：资源部分的一个子集，通常用于表示数据表中的一条记录。如果操作的是整个资源(数据表)，该部分可省略。

以下协议是一个标准 URI，用于操作“user”表中 ID 为“1”的记录：

```
content://com.yg.providers.myprovider/user/1
```

其中：“com.yg.providers.myprovider”对应授权部分；“user/1”对应资源部分。

Content Provider 使用的 Uri 协议中，协议头和权限部分是不可或缺的。

1.2　系统通讯录

Android API 将系统通讯录以 Content Provider 的方式提供给开发者，使其可以对通讯录进行一系列的操作。

1.2.1　系统通讯录结构

Android 系统通过一个单独的 Content Provider 程序来向其他程序提供系统通讯录相关操作。想要了解系统通讯录数据在数据库中存储的结构，需要将此程序的数据库文件导出，之后通过 SQLite 数据库管理工具进行查看。

自 Android4.0 起，DDMS 程序已经无法查看手机中已安装程序的数据，但可以查看模拟器中的程序信息。因此，接下来将使用模拟器进行导出操作。

在“DDMS”界面中，打开“File Explorer”文件浏览器窗口，系统通讯录 Content Provider 程序位于“data/data/com.android.providers.contacts”目录下，数据保存在该目录中的“databases/contacts2.db”数据库文件中，将其导出，并使用 SQLite 数据库管理工具打开。

系统通讯录数据库中存在若干张数据表，但与本小节相关的表只有三张：“mimetypes”表、“raw_contacts”表和“data”表。

1. mimetypes 表

“mimetypes”表用于存放数据类型，这些数据类型用于区分“data”表中的数据。

Android 提供了相应的常量与“mimetypes”表中列出的数据类型相对应，这些常量均由CommonDataKinds 类中的内部类描述，CommonDataKinds 类中常用内部类如表 1-2 所示。

表 1-2 CommonDataKinds 类中常用内部类

类 名	描 述
StructuredName	描述姓名类型数据
Phone	描述电话号码类型数据
Email	描述电子邮箱类型数据
Photo	描述头像类型数据
StructuredPostal	描述联系地址类型数据

表 1-2 中的类分别描述了各种数据类型和对应表中的字段名等常量。其中，这些类有一个共同的常量“CONTENT_ITEM_TYPE”，对应“mimetypes”表中列出的数据类型，对应关系如表 1-3 所示。

表 1-3 “mimetypes”表中常用数据类型对应常量

数 据 类 型	常 量
vnd.android.cursor.item/name	StructuredName.CONTENT_ITEM_TYPE
vnd.android.cursor.item/phone_v2	Phone.CONTENT_ITEM_TYPE
vnd.android.cursor.item/email_v2	Email.CONTENT_ITEM_TYPE
vnd.android.cursor.item/photo	Photo.CONTENT_ITEM_TYPE
vnd.android.cursor.item/postal-address_v2	StructuredPostal.CONTENT_ITEM_TYPE

2. raw_contacts 表

“raw_contacts”表是通讯录的主表，用于存放联系人的详细数据，但通常用于操作的表是“data”表，在“data”表中添加新联系人的信息之前，必须在“raw_contacts”表中生成新联系人的 ID。

操作“raw_contacts”表的 Uri 是“content://com.android.contacts/raw_contacts”，对应的常量为：RawContacts.CONTENT_URI。

3. data 表

“data”表用于存放具体联系人的数据，是“raw_contacts”表的子表。其主要有三个字段：“raw_contact_id”、“data1”和“mimetype_id”。

✧ “raw_contact_id”描述联系人唯一标识，与“raw_contacts”表相关联。

✧ “data1”描述联系人资料，例如：姓名、电话号码、电子邮箱等信息。

✧ “mimetype_id”与“mimetypes”表相关联，用于描述“data1”中数据的类型。

在系统通讯录中添加几条联系人数据后，导出数据库，“data”表中主要列的内容如图 1-2 所示。

观察图 1-2 不难发现，联系人的不同信息都被保存到了“data1”字段中，这些不同的信息用“mimetype_id”来区分。因此，如果要对联系人信息进行操作，对象主要是“data”表中的“data1”字段。

操作“data”表的 Uri 是“content://com.android.contacts/data”，对应的常量为：

ContactsContract.Data.CONTENT_URI。

mimetype_id	raw_contact_id	data1
(empty)	(empty)	(empty)
5	1	1 562-889-1233
8	1	Shandong Qingdao
1	1	lilei@mail.com
7	1	LiLei
5	2	1 866-192-9369
8	2	Beijing
1	2	tom@mail.com
7	2	Tom
5	3	1 585-678-8959
8	3	Shandong Jinan
1	3	lucy@mail.com
7	3	Lucy

图 1-2　“data”表中主要列的内容

1.2.2　操作系统通讯录

在了解了系统通讯录数据组织结构后，本小节将编写程序，对系统通讯录进行操作。

下述示例用于实现：利用系统提供的通讯录 Content Provider，对通讯录进行“增删改查”操作。要求如下：

✧ 打开程序，查询通讯录中所有联系人，并显示到界面中。

✧ 添加新的联系人到通讯录。

✧ 修改、删除通讯录中的联系人。

对通讯录进行操作的步骤如下：

(1) 创建项目“ch01_ContactManager”，首先创建一个联系人实体类，用于封装联系人信息，包括联系人 ID、名称和电话号码等属性。在“com.yg.ch01_contactmanager.entity”包中创建“Contact.java”类，代码如下：

```
public class Contact {
    private String id;
    private String name;
    private String phone;

  // 属性 get set 方法略
    @Override
    public String toString() {
        return "Contact [id=" + id + ", name=" + name + ", phone="
                + phone + "]";
    }
}
```

(2) 编写工具类，对系统通讯录进行“增删改查”等操作。创建“ContactHelper.java”类，该类是项目中的核心类。首先添加 addContact()方法，代码如下：

```
public class ContactHelper {
    /**
     * 添加一个联系人到通讯录
     */
    public static boolean addContact(Context ctx, Contact contact) {
        // 获取联系人信息
        String name = contact.getName();
        String phone = contact.getPhone();
        ContentResolver resolver = ctx.getContentResolver();
        ContentValues values = new ContentValues();
        Uri uri = null;
        long newId = 0;

        // 第一步：在 raw_contacts 表中插入一条空记录，获取返回的 ID 值
        uri = RawContacts.CONTENT_URI;
        // 插入数据，返回 Uri 对象
        Uri rstUri = resolver.insert(uri, values);
        // 解析该 Uri 对象，获取新记录的 ID
        newId = ContentUris.parseId(rstUri);

        // 第二步：插入联系人姓名
        uri = Data.CONTENT_URI;
        values.put(Data.RAW_CONTACT_ID, newId);
        values.put(StructuredName.DISPLAY_NAME, name);
        values.put(Data.MIMETYPE, StructuredName.CONTENT_ITEM_TYPE);
        resolver.insert(uri, values);

        // 第三步：
        values.clear();
        values.put(Data.RAW_CONTACT_ID, newId);
        values.put(Phone.NUMBER, phone);
        values.put(Data.MIMETYPE, Phone.CONTENT_ITEM_TYPE);
        resolver.insert(uri, values);

        return true;
    }
}
```

该方法用于把传入的 Contact 联系人对象与系统通讯录表中的字段对应起来，封装为一个 ContentValues 对象，之后将联系人添加到系统通讯录中。向通讯录中插入联系人的步骤如下：

(1) 在通讯录中插入空记录，解析返回的 Uri 对象，得到记录 ID；

(2) 使用获取到的“记录 ID”向通讯录添加联系人姓名；

(3) 使用获取到的“记录 ID”向通讯录添加联系人电话号码。

注 意　由于“data”表存储联系人信息的方式比较特殊，因此，同一个联系人的信息通常会分多次添加或修改。

接下来编写“删除联系人”的方法，代码如下：

```
/**
 * 根据联系人 id 删除联系人
 */
public static boolean deleteContact(Context ctx, String contactId) {
        ContentResolver resolver = ctx.getContentResolver();
        // 删除 raw_contacts 表中的记录
        int count = resolver.delete(RawContacts.CONTENT_URI, Contacts._ID
                        + "=?", new String[] { contactId });
        return count > 0;
}
```

该方法通过传入的联系人 ID 删除系统通讯录中对应的联系人，如果返回值“count”大于 0，则表示删除成功。

继续编写“更新联系人”的方法，代码如下：

```
/**
 * 更新联系人
 */
public static boolean updateContact(Context ctx, Contact contact) {
        // 获取联系人信息
        String contactId = contact.getId();
        String name = contact.getName();
        String phone = contact.getPhone();
        ContentResolver resolver = ctx.getContentResolver();
        ContentValues values = new ContentValues();
        Uri uri = null;

        // 更新条件
        String selection = Data.MIMETYPE + "=? and " + Data.RAW_CONTACT_ID
                        + "=?";
        String[] selectionArgs = null;

        // 更新联系人姓名
        uri = Data.CONTENT_URI;
        // 往 data 表中插入姓名数据
```

```
        values.put(StructuredName.GIVEN_NAME, name);
        selectionArgs = new String[] { StructuredName.CONTENT_ITEM_TYPE,
                        contactId };
        int count = resolver.update(uri, values, selection, selectionArgs);

        values.clear();
        values.put(Phone.NUMBER, phone);
        selectionArgs = new String[] { Phone.CONTENT_ITEM_TYPE, contactId };
        count = resolver.update(uri, values, selection, selectionArgs);

        return count > 0;
}
```

该方法通过联系人的 ID，更新联系人信息，如果返回值“count”大于 0，则表示更新成功。

继续编写“获取手机联系人”的方法，代码如下：

```
/**
 * 获取手机联系人
  */
public static List<Contact> getContacts(Context ctx) {

        List<Contact> contacts = new ArrayList<Contact>();
        ContentResolver resolver = ctx.getContentResolver();
        // 查询联系人
        Cursor cursor = resolver.query(Contacts.CONTENT_URI, null, null, null,
                        null);

        // 获取 ID 字段的索引
        int idIndex = cursor.getColumnIndex(Contacts._ID);
        // 获取联系人姓名字段索引
        int nameIndex = cursor.getColumnIndex(Contacts.DISPLAY_NAME);

        while (cursor.moveToNext()) {
                Contact contact = new Contact();
                // 获取联系人 id
                String contactId = cursor.getString(idIndex);
                // 获取联系人姓名
                String name = cursor.getString(nameIndex);
                contact.setId(contactId);
                contact.setName(name);
```

```
            // 根据联系人 ID 查询对应的电话号码
            String selection = Data.MIMETYPE + "=? and "
                    + Data.RAW_CONTACT_ID + "=?";
            String[] selectionArgs = new String[] { Phone.CONTENT_ITEM_TYPE,
                    contactId };

            Cursor phoneNumsCursor = resolver.query(Data.CONTENT_URI, null,
                    selection, selectionArgs, null);
            // 获取电话号，可能存在多个，在此只取第一个
            if (phoneNumsCursor.moveToNext()) {
                String phone = phoneNumsCursor.getString(phoneNumsCursor
                        .getColumnIndex(Phone.NUMBER));
                contact.setPhone(phone);
            }
            phoneNumsCursor.close();
            contacts.add(contact);
        }
        cursor.close();
        return contacts;
}
```

该方法中，首先通过执行 ContentResolver.query()方法获得联系人数据表的 Cursor 游标对象，之后遍历这个游标，将联系人添加到 List<Contact>集合中，并返回。

至此，工具类“ContactHelper.java”编写完成。

(3) 为联系人列表创建适配器，首先编写适配器中 Item 的布局文件。创建“item_main_list.xml”布局文件，代码如下：

```
<RelativeLayout xmlns:android="http://schemas.android.com/apk/res/android"
    xmlns:tools="http://schemas.android.com/tools"
    android:layout_width="match_parent"
    android:layout_height="match_parent"
    android:descendantFocusability="blocksDescendants"
    android:orientation="horizontal"
    android:padding="10dp" >

    <LinearLayout
        android:layout_width="wrap_content"
        android:layout_height="wrap_content"
        android:layout_centerVertical="true"
        android:orientation="vertical" >
        <TextView
            android:id="@+id/item_main_list_name_tv"
```

```
            android:layout_width="wrap_content"
            android:layout_height="wrap_content" />
        <TextView
            android:id="@+id/item_main_list_phone_tv"
            android:layout_width="wrap_content"
            android:layout_height="wrap_content"
            android:layout_marginTop="5dp" />
    </LinearLayout>

    <Button
        android:id="@+id/item_main_list_del_btn"
        android:layout_width="wrap_content"
        android:layout_height="wrap_content"
        android:layout_alignParentRight="true"
        android:layout_centerVertical="true"
        android:text="删除" />
</RelativeLayout>
```

然后在“com.yg.ch01_contactmanager.adapter”包中创建“ContactListAdapter.java”适配器类，代码如下：

```
public class ContactListAdapter extends BaseAdapter {
    private List<Contact> contacts = null;
    private Context ctx = null;
    public ContactListAdapter(Context ctx, List<Contact> contacts) {
        this.ctx = ctx;
        this.contacts = contacts;
    }
    @Override
    public int getCount() {
        return contacts.size();
    }
    @Override
    public Object getItem(int position) {
        return contacts.get(position);
    }
    @Override
    public long getItemId(int position) {
        return 0;
    }
    @Override
    public View getView(int position,View convertView,ViewGroup parent) {
```

```
            if (convertView == null) {
                convertView = LayoutInflater.from(ctx).inflate(
                                R.layout.item_main_list, null);
            }
            final Contact contact = contacts.get(position);

            TextView nameTv = (TextView) convertView
                        .findViewById(R.id.item_main_list_name_tv);
            TextView phoneTv = (TextView) convertView
                        .findViewById(R.id.item_main_list_phone_tv);
            Button delBtn = (Button) convertView
                        .findViewById(R.id.item_main_list_del_btn);

            nameTv.setText(contact.getName());
            phoneTv.setText(contact.getPhone());
            delBtn.setOnClickListener(new OnClickListener() {

                @Override
                public void onClick(View view) {
                    new AlertDialog.Builder(ctx)
                            .setMessage("确定要删除联系人吗？")
                            .setNeutralButton("取消", null)
                            .setPositiveButton("确定",
                                new DialogInterface.OnClickListener() {
                                    @Override
                                    public void onClick(DialogInterface dialog,
                                        int which) {
                                        // 删除联系人
                                        deleteContact(contact);
                                    }
                                }).show();
                }
            });
            return convertView;
        }

        /**
         * 删除联系人
         */
        private void deleteContact(Contact contact) {
```

```
                boolean flag = ContactHelper.deleteContact(ctx, contact.getId());
                if (flag) {
                        contacts.remove(contact);
                        this.notifyDataSetChanged();
                }
        }
}
```

该适配器中，为每个 Item 添加了一个“删除”按钮，当点击“删除”按钮时，系统通讯录中对应的联系人会被删除。

(4) 实现主界面功能，该功能主要用于显示联系人列表，添加和修改联系人。首先修改“activity_main.xml”布局文件，代码如下：

```
<RelativeLayout xmlns:android="http://schemas.android.com/apk/res/android"
    xmlns:tools="http://schemas.android.com/tools"
    android:layout_width="match_parent"
    android:layout_height="match_parent" >
    <ListView
        android:id="@+id/act_main_listview"
        android:layout_width="match_parent"
        android:layout_height="match_parent"
        android:layout_above="@+id/act_main_add_btn" >
    </ListView>
    <Button
        android:id="@+id/act_main_add_btn"
        android:layout_width="match_parent"
        android:layout_height="wrap_content"
        android:text="添加"
        android:layout_alignParentBottom="true" />
</RelativeLayout>
```

添加或修改联系人信息的功能是通过 AlertDialog 提示框的形式实现的，因此，需要为提示框添加自定义的 View 视图。创建 AlertDialog 自定义视图布局文件“dialog_contact_info_layout.xml”，代码如下：

```
<LinearLayout xmlns:android="http://schemas.android.com/apk/res/android"
    xmlns:tools="http://schemas.android.com/tools"
    android:layout_width="match_parent"
    android:layout_height="match_parent"
    android:orientation="vertical"
     android:padding="10dp" >

    <LinearLayout
        android:layout_width="fill_parent"
```

```
        android:layout_height="wrap_content"
        android:orientation="horizontal"
        android:padding="10dp" >
        <TextView
            android:layout_width="wrap_content"
            android:layout_height="wrap_content"
            android:text="联　系　人" />
        <EditText
            android:id="@+id/dialog_contact_name_et"
            android:layout_width="fill_parent"
            android:layout_height="wrap_content"
            android:layout_marginLeft="10dp" />
    </LinearLayout>
    <LinearLayout
        android:layout_width="fill_parent"
        android:layout_height="wrap_content"
        android:orientation="horizontal"
        android:padding="10dp" >
        <TextView
            android:layout_width="wrap_content"
            android:layout_height="wrap_content"
            android:text="联系方式" />
        <EditText
            android:id="@+id/dialog_contact_phone_et"
            android:layout_width="fill_parent"
            android:layout_height="wrap_content"
            android:layout_marginLeft="10dp"
            android:inputType="phone" />
    </LinearLayout>
</LinearLayout>
```

最后实现主界面功能。分步骤修改“MainActivity.java”类，基本代码如下：

```
public class MainActivity extends Activity {
        private ListView listView = null;
        private ContactListAdapter adapter = null;
        private List<Contact> contacts = new ArrayList<Contact>();

        @Override
        protected void onCreate(Bundle savedInstanceState) {
                super.onCreate(savedInstanceState);
                setContentView(R.layout.activity_main);
```

```
            listView = (ListView) findViewById(R.id.act_main_listview);
            adapter = new ContactListAdapter(this, contacts);
            listView.setAdapter(adapter);
            listView.setOnItemClickListener(new OnItemClickListener() {
                    @Override
                    public void onItemClick(AdapterView<?> parent, View view,
                            int position, long id) {
                            final Contact contact = contacts.get(position);
                            showDialog(contact);
                    }
            });

            Button addBtn = (Button) findViewById(R.id.act_main_add_btn);
            addBtn.setOnClickListener(new OnClickListener() {
                    @Override
                    public void onClick(View arg0) {
                            showDialog(null);
                    }
            });
            // 显示联系人列表
            showContactList();
    }
}
```

上述代码仅实现了基本功能，包括控件的引用以及事件的添加。接下来添加showContactList()方法，用于获取系统通讯录列表，并显示到界面中，代码如下：

```
/**
 * 显示联系人列表
 */
private void showContactList() {
        List<Contact> contactsTemp = ContactHelper.getContacts(this);
        contacts.removeAll(contacts);
        contacts.addAll(contactsTemp);
        adapter.notifyDataSetChanged();
}
```

添加或修改联系人的功能用同一个 AlertDialog 对话框来实现。显示对话框的代码如下：

```
/**
 * 显示对话框，用于添加或修改联系人
 */
private void showDialog(final Contact contact) {
```

```
        View view = LayoutInflater.from(MainActivity.this).inflate(
                        R.layout.dialog_contact_info_layout, null);

        final EditText nameEt = (EditText) view
                        .findViewById(R.id.dialog_contact_name_et);
        final EditText phoneEt = (EditText) view
                        .findViewById(R.id.dialog_contact_phone_et);

        // 如果 contact 不为空，则说明当前为更新联系人，否则为添加新联系人
        if (contact != null) {
                nameEt.setText(contact.getName());
                phoneEt.setText(contact.getPhone());
        }

        new AlertDialog.Builder(MainActivity.this).setView(view)
                .setCancelable(false).setNegativeButton("取消", null)
                .setPositiveButton("确定",
                        new DialogInterface.OnClickListener() {
                            @Override
                            public void onClick(DialogInterface dialog,int which){
                                    String name = nameEt.getText().toString();
                                    String phone = phoneEt.getText().toString();

                                    // 判断是添加联系人，还是更新联系人
                                    if (contact == null) {
                                            Contact ctt = new Contact();
                                            ctt.setName(name);
                                            ctt.setPhone(phone);
                                            // 添加联系人
                                            addContact(ctt);
                                    } else {
                                            contact.setName(name);
                                            contact.setPhone(phone);
                                            // 更新联系人
                                            updateContact(contact);
                                    }

                            }
                        }).show();
}
```

该方法中，如果传入的 Contact 对象为“null”，则表示当前是添加联系人操作，否则为修改联系人操作。当点击“确定”按钮后，根据操作目的不同，调用不同的方法。其中addContact()方法用于添加联系人，updateContact()方法用于更新联系人信息。这两个方法代码如下：

```
/**
 * 添加联系人
 */
protected void addContact(Contact contact) {
        ContactHelper.addContact(MainActivity.this, contact);
        showContactList();
}
/**
 * 更新联系人
 */
protected void updateContact(Contact contact) {
        ContactHelper.updateContact(MainActivity.this, contact);
        showContactList();
}
```

当对联系人列表进行了修改后，重新执行 showContactList()方法，获取最新的联系人列表。至此，主界面功能已实现。

(5) 最后，在“AndroidManifest.xml”配置文件中，添加读写系统通讯录的相关权限，代码如下：

```
<uses-permission android:name="android.permission.READ_CONTACTS" />
<uses-permission android:name="android.permission.WRITE_CONTACTS" />
```

(6) 运行程序后，效果如图 1-3 所示。

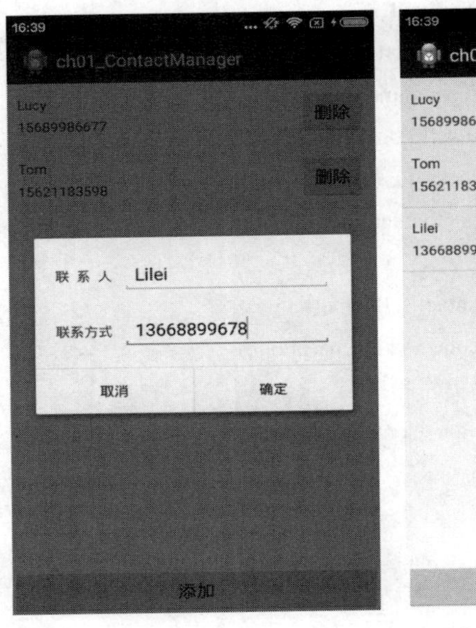

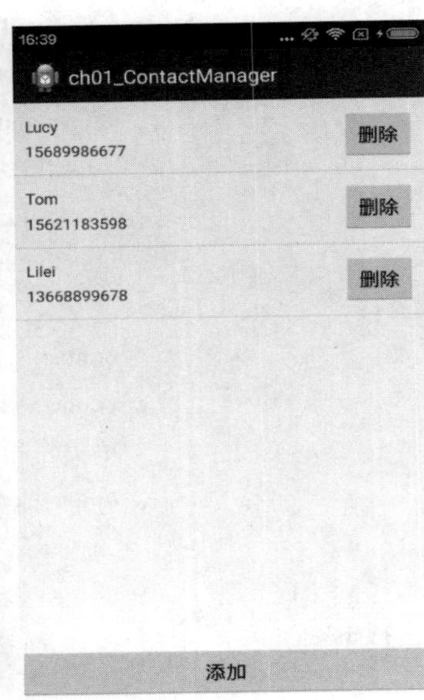

图 1-3　通讯录程序界面

1.3　自定义 Content Provider

在实际开发中，开发者可能打算将所开发程序中的部分或全部数据共享给其他程序使用，此时，就可以通过自定义 Content Provider 的方式来解决。

程序中可以实现 Content Provider 功能，为了方便数据的存储与管理，通常使用 SQLite 数据库作为数据的载体。本小节将介绍如何自定义 Content Provider 程序，以及如何使用 Content Provider 程序操作数据。

1.3.1　创建 Content Provider

下述示例用于实现：以 SQLite 数据库作为数据源，创建 Content Provider 程序，对其他程序提供简单的用户管理功能，并能够对用户进行“增删改查”等操作。

(1) 创建项目“ch01_MyContentProvider”。首先创建数据库，新建“DBHelper.java”类，代码如下：

```
public class DBHelper extends SQLiteOpenHelper {

	private static final int DB_VERSION = 1;
	private static final String DB_NAME = "my_database";

	public static final String TABLE_NAME_USERS = "users";
	public static final String COLUMN_ID = "_id";

	public DBHelper(Context context) {
		super(context, DB_NAME, null, DB_VERSION);
	}

	@Override
	public void onCreate(SQLiteDatabase db) {
		// 创建数据表
		String createTableSql = " create table " + TABLE_NAME_USERS
				+ " ("
				+ COLUMN_ID + " integer primary key autoincrement,"
				+ " name text,age integer)";
		db.execSQL(createTableSql);
	}

	@Override
```

```
public void onUpgrade(SQLiteDatabase db, int oldVersion,
        int newVersion){ }
/**
 * 插入一条用户信息
 */
public int insertUser(String name, int age) {
        SQLiteDatabase db = getWritableDatabase();

        String sql = "insert into " + DBHelper.TABLE_NAME_USERS
                    + "(name,age) values(?,?)";
        db.execSQL(sql, new Object[] { name, age });

        Cursor cursor = db.rawQuery("select last_insert_rowid() from "
                    + DBHelper.TABLE_NAME_USERS, null);
        int id = -1;
        if (cursor.moveToFirst())
                id = cursor.getInt(0);
        return id;
}
/**
 * 删除指定用户信息
 */
public int deleteUserById(String id) {
        SQLiteDatabase db = getWritableDatabase();
        int count = db.delete(DBHelper.TABLE_NAME_USERS,
                DBHelper.COLUMN_ID + "=?", new String[] { id + "" });
        return count;
}
/**
 * 删除所有用户信息
 */
public int deleteAllUser() {
        SQLiteDatabase db = getWritableDatabase();
        int count = db.delete(DBHelper.TABLE_NAME_USERS, null, null);
        return count;
}
/**
 * 更新指定用户信息
 */
public int updateUserById(String id, ContentValues values) {
```

```
            SQLiteDatabase db = getWritableDatabase();
            int count = db.update(DBHelper.TABLE_NAME_USERS, values,
                        DBHelper.COLUMN_ID + "=?", new String[] { id });
            return count;
    }
    /**
     * 查找所有用户信息
     */
    public Cursor findAllUser() {
            SQLiteDatabase db = getReadableDatabase();
            Cursor cursor = db.query(TABLE_NAME_USERS, null, null, null,
                    null, null, null);
            return cursor;
    }

}
```

上述代码，主要用于创建一个数据库。在数据库中创建了一个“users”表，并在该类中实现对该表的“增删改查”操作。

注 意　关于数据库的知识，本节不做过多讲解，需要注意的是：在编写 SQL 语句时，一定不要忘记插入该有的空格，尤其是拼接语句，例如本例中创建“users”表的语句。

(2) 创建 Content Provider。首先创建“MyProvider.java”类，并继承“ContentProvider”抽象类，实现必要的抽象方法，代码如下：

```
public class MyProvider extends ContentProvider {

    private static final UriMatcher matcher;

    private static final int ALL_USER = 1;
    private static final int SINGLE_USER = 2;

    private DBHelper dbHelper = null;
    static {
            // 创建 UriMatcher 对象
            matcher = new UriMatcher(UriMatcher.NO_MATCH);
            // 向 UriMatcher 对象添加规则
            // users 结尾的 uri 表示操作所有数据
            matcher.addURI("com.yg.provider.userprovider",
                    "users", ALL_USER);
            // users/# 结尾的 uri 表示操作某条数据，其中“#”表示 id 值
            matcher.addURI("com.yg.provider.userprovider", "users/#",
```

```
                SINGLE_USER);
    }

    @Override
    public boolean onCreate() {
        // 在 onCreate()中，创建数据库实例
        dbHelper = new DBHelper(getContext());
        return true;
    }

    @Override
    public Uri insert(Uri uri, ContentValues values) {

        //匹配 uri
        if (matcher.match(uri) == ALL_USER) {
            String name = values.getAsString("name");
            Integer age = values.getAsInteger("age");

            if (TextUtils.isEmpty(name) || age == null) {
                return null;
            }

            SQLiteDatabase db = dbHelper.getWritableDatabase();

            String sql = "insert into " + DBHelper.TABLE_NAME_USERS
                    + "(name,age) values(?,?)";
            db.execSQL(sql, new Object[] { name, age });

            Cursor cursor = db.rawQuery("select last_insert_rowid()
                from " + DBHelper.TABLE_NAME_USERS, null);
            int id = -1;
            if (cursor.moveToFirst())
                id = cursor.getInt(0);

            Uri newUri = ContentUris.withAppendedId(uri, id);
            return newUri;
        }

        return null;
    }
```

```
@Override
public int update(Uri uri, ContentValues values, String where,
            String[] whereArgs) {
        SQLiteDatabase db = dbHelper.getReadableDatabase();
        int count = 0;
        //匹配 uri
        switch (matcher.match(uri)) {
        case ALL_USER:
                count = db.update(DBHelper.TABLE_NAME_USERS, values, where,
                        whereArgs);
                break;

        case SINGLE_USER:
                String id = uri.getPathSegments().get(1);

                String selection = "";
                if (!TextUtils.isEmpty(where)) {
                        selection = " AND (" + where + ")";
                }
                count = db.update(DBHelper.TABLE_NAME_USERS, values,
                        DBHelper.COLUMN_ID + "=" + id + selection, whereArgs);
                break;
        default:
                return count;
        }

        return count;
}

@Override
public int delete(Uri uri, String where, String[] whereArgs) {
        int count = 0;
        SQLiteDatabase db = dbHelper.getWritableDatabase();
        System.out.println("del:" + uri);
        //匹配 uri
        switch (matcher.match(uri)) {
        case ALL_USER:
                count = db.delete(DBHelper.TABLE_NAME_USERS, where,
                        whereArgs);
```

```
                break;
        case SINGLE_USER:
                String id = uri.getPathSegments().get(1);

                String selection = "";
                if (!TextUtils.isEmpty(where)) {
                        selection = " AND (" + where + ")";
                }
                count = db.delete(DBHelper.TABLE_NAME_USERS,
                        DBHelper.COLUMN_ID + "=" + id + selection, whereArgs);
                break;
        }

        return count;
}

@Override
public Cursor query(Uri uri, String[] projection, String selection,
                String[] selectionArgs, String sortOrder) {

        SQLiteQueryBuilder qb = new SQLiteQueryBuilder();
        qb.setTables(DBHelper.TABLE_NAME_USERS);
        //匹配 uri
        switch (matcher.match(uri)) {
        case ALL_USER:
                break;

        case SINGLE_USER:
                qb.appendWhere(DBHelper.COLUMN_ID + "="
                                + uri.getPathSegments().get(1));
                break;
        default:
                return null;
        }
        SQLiteDatabase db = dbHelper.getReadableDatabase();
        Cursor c = qb.query(db, projection, selection, selectionArgs,
                null, null, sortOrder);
        return c;
}
```

```
    @Override
    public String getType(Uri uri) {
        return null;
    }
}
```

该类继承了 Content Provider 类，并实现了必要的方法。具体方法解析如下：

- ✧ 在 static 静态代码块中，创建 UriMatcher 对象，并添加 Uri 匹配规则，本程序基本 Uri 为“com.yg.provider.userprovider”。如果以“users”结尾，则表示操作所有数据；如果以“users/#”结尾，则表示操作单条数据，例如以“users/1”结尾，表示操作 ID 值为 1 的记录。
- ✧ onCreate()方法与 Activity 中的 onCreate()方法类似，只执行一次，在此方法中，实例化数据库对象。
- ✧ insert()方法用于插入数据，代码中，只针对操作所有数据的 Uri 进行匹配(“users”结尾的 Uri)，获取“name”和“age”数据，并把它们插入到数据库中，随后获取当前插入数据的 ID 值，通过“ContentUris.withAppendedId()”生成带有 ID 结尾的 Uri，并返回。
- ✧ update()方法用于更新数据，代码中，先要对 Uri 进行匹配，如果操作所有数据，则直接执行更新数据库的方法；如果操作指定数据，则首先获取 ID，然后构建更新条件语句，再执行更新方法，此方法返回值为更新记录的条数。
- ✧ delete()方法用于删除数据，类似于 update()方法。此方法返回值为删除记录的条数。
- ✧ query()方法用于查询数据，此方法使用 SQLiteQueryBuilder 类对数据进行查询，先要与 Uri 匹配，根据匹配结果来制定查询条件并执行查询操作，返回值为查询后的 Cursor 对象。
- ✧ getType()方法用于获取 MIME 类型，本程序未涉及，因此，此方法为空。

(3) 注册 Content Provider。完成 Content Provider 的创建之后，使用前，必须同 Activity 与 Service 一样，在 AndroidManifest 配置文件中进行注册，具体代码如下所示：

```
<provider android:name=".MyProvider"
        android:authorities="com.yg.provider.userprovider" />
```

通过<provider>标签来注册 Content Provider，属性介绍如下：

- ✧ name：声明 ContentProvider 类的路径。
- ✧ authorities：声明 ContentProvider 的基本 Uri，就像 Android 程序的包名一样重要，必须保证唯一。因此建议使用当前程序的包名作为前缀，例如本示例中，程序包名为“com.yg.provider”，声明的 authorities 为“com.yg.provider.userprovider”。

1.3.2 使用自定义的 Content Provider

本小节将利用在 1.3.1 小节中创建的自定义 Content Provider 实现对用户的“增删改

查”操作。简单起见，本小节示例不再创建新项目，将直接在 1.3.1 小节项目基础上进行操作。程序要求如下：

- ✧ 主界面显示用户信息列表。
- ✧ 主界面提供添加用户按钮。
- ✧ 点击列表项，修改用户信息。
- ✧ 长按列表项，删除用户。

使用自定义的 Content Provider 实现对用户的“增删改查”，操作步骤如下：

(1) 在“com.yg.provider.test”包中创建用户实体类“User.java”，代码如下：

```
public class User {
	private int id;
	private String name;
	private int age;

	public User(String name, int age) {
		super();
		this.name = name;
		this.age = age;
	}
	public User(int id, String name, int age) {
		this.id = id;
		this.name = name;
		this.age = age;
	}
	//属性 get set  方法略
	@Override
	public String toString() {
		return "姓名：" + name + "  年龄：" + age;
	}
}
```

(2) 在“com.yg.provider.test”包中创建工具类“UserHelper.java”，该类提供了操作“用户信息 Content Provider”的相关方法，代码如下：

```
public class UserHelper {
	private static final Uri USER_PROVIDER_URI = Uri
			.parse("content://com.yg.provider.userprovider/users");
	private static final String USER_ID = "_id";
	private static final String USER_NAME = "name";
	private static final String USER_AGE = "age";

	/**
	 * 添加用户
```

```
     */
    public static boolean addUser(Context ctx, User user) {
            ContentResolver resolver = ctx.getContentResolver();
            ContentValues values = new ContentValues();
            values.put(USER_NAME, user.getName());
            values.put(USER_AGE, user.getAge());
            Uri reqUri = resolver.insert(USER_PROVIDER_URI, values);

            List<String> reqs = reqUri.getPathSegments();
            return reqs == null || reqs.size() == 0;
    }
    /**
     * 删除用户
     */
    public static boolean delUser(Context ctx, int userId) {
            ContentResolver resolver = ctx.getContentResolver();
            Uri delUri = ContentUris.withAppendedId(USER_PROVIDER_URI,
                    userId);
            int count = resolver.delete(delUri, null, null);
            return count > 0;
    }
    /**
     * 更新用户
     */
    public static boolean updateUser(Context ctx, User user) {
            ContentResolver resolver = ctx.getContentResolver();
            ContentValues values = new ContentValues();
            values.put(USER_NAME, user.getName());
            values.put(USER_AGE, user.getAge());
            Uri updateUri = ContentUris.withAppendedId(USER_PROVIDER_URI,
                            user.getId());
            int count = resolver.update(updateUri, values, null, null);
            return count > 0;
    }
    /**
     * 查询用户
     */
    public static List<User> getUsers(Context ctx) {
            List<User> users = new ArrayList<User>();
            ContentResolver resolver = ctx.getContentResolver();
```

```
            Cursor cursor = resolver.query(USER_PROVIDER_URI, null, null,   null, null);
            if (cursor == null) {
                return null;
            }
            while (cursor.moveToNext()) {
                int columnid = cursor.getColumnIndex(USER_ID);
                int columnName = cursor.getColumnIndex(USER_NAME);
                int columeAge = cursor.getColumnIndex(USER_AGE);
                int id = cursor.getInt(columnid);
                String name = cursor.getString(columnName);
                int age = cursor.getInt(columeAge);

                User u = new User(id, name, age);
                users.add(u);
            }
            return users;
    }
}
```

上述代码实现 Content Provider 的“增删改查”功能的方式与 1.2.2 小节中操作系统通讯录的方式类似。

(3) 完成主界面的编写，首先编写主界面布局文件“activity_main.xml”，代码如下：

```
<RelativeLayout xmlns:android="http://schemas.android.com/apk/res/android"
    xmlns:tools="http://schemas.android.com/tools"
    android:layout_width="match_parent"
    android:layout_height="match_parent" >

    <ListView
        android:id="@+id/act_main_listview"
        android:layout_width="match_parent"
        android:layout_height="match_parent"
        android:layout_above="@+id/act_main_add_btn" >
    </ListView>
    <Button
        android:id="@+id/act_main_add_btn"
        android:layout_width="match_parent"
        android:layout_height="wrap_content"
        android:layout_alignParentBottom="true"
        android:text="添加" />
</RelativeLayout>
```

(4) 将已有的“MainActivity.java”类移动到“com.yg.provider.test”包中，修改代码

如下：

```
public class MainActivity extends Activity {
        private Context ctx = null;
        private Button addBtn = null;
        private ListView listView = null;
        private ArrayAdapter<User> adapter = null;
        private List<User> users = new ArrayList<User>();

        @Override
        protected void onCreate(Bundle savedInstanceState) {
                super.onCreate(savedInstanceState);
                setContentView(R.layout.activity_main);

                ctx = this;
                addBtn = (Button) findViewById(R.id.act_main_add_btn);
                listView = (ListView) findViewById(R.id.act_main_listview);
                adapter = new ArrayAdapter<User>(this,
                        android.R.layout.simple_list_item_1, users);
                listView.setAdapter(adapter);

                addBtn.setOnClickListener(new OnClickListener() {
                        @Override
                        public void onClick(View view) {
                                showDialog(null);
                        }
                });
                listView.setOnItemClickListener(new OnItemClickListener() {
                        @Override
                        public void onItemClick(AdapterView<?> parent, View view,
                                        int position, long id) {
                                User user = users.get(position);
                                showDialog(user);
                        }
                });
                listView.setOnItemLongClickListener(
                    new OnItemLongClickListener(){
                        @Override
                        public boolean onItemLongClick(AdapterView<?> parent,
                            View view, int position, long id) {
                                User user = users.get(position);
```

```
                    boolean flag = UserHelper.delUser(ctx, user.getId());
                    Toast.makeText(ctx, flag ? "删除成功" : "删除失败",
                            Toast.LENGTH_SHORT).show();
                    showUsers();
                    return false;
                }
            });
            showUsers();
    }
    /**
     * 显示对话框，用于添加或修改用户信息
     *
     * @param contact
     */
    private void showDialog(final User user) {
            View view = LayoutInflater.from(MainActivity.this).inflate(
                            R.layout.dialog_user_info_layout, null);

            final EditText nameEt = (EditText) view
                            .findViewById(R.id.dialog_user_name_et);
            final EditText ageEt = (EditText) view
                            .findViewById(R.id.dialog_user_age_et);

            // 如果 contact 不为空，则说明当前为更新用户信息，否则为添加新用户
            if (user != null) {
                    nameEt.setText(user.getName());
                    ageEt.setText(user.getAge() + "");
            }
            new AlertDialog.Builder(MainActivity.this).setView(view)
                    .setCancelable(false).setNegativeButton("取消", null)
                    .setPositiveButton("确定",
                        new DialogInterface.OnClickListener() {
                            @Override
                            public void onClick(DialogInterface dialog,
                                    int which){
                                    String name = nameEt.getText().toString();
                                    String t = ageEt.getText().toString();
                                    int age = Integer.parseInt(t);

                                    if (user == null) {
```

```
                        User u = new User(name, age);
                        // 添加新用户
                        UserHelper.addUser(ctx, u);
                    } else
                    {
                        user.setName(name);
                        user.setAge(age);
                        // 更新用户信息
                        UserHelper.updateUser(ctx, user);
                    }
                    showUsers();
                }
            }).show();
    }
    private void showUsers() {
        users.removeAll(users);
        users.addAll(UserHelper.getUsers(this));
        adapter.notifyDataSetChanged();
    }
}
```

至此，主界面的编写已完成。最后运行程序，效果如图 1-4 所示。

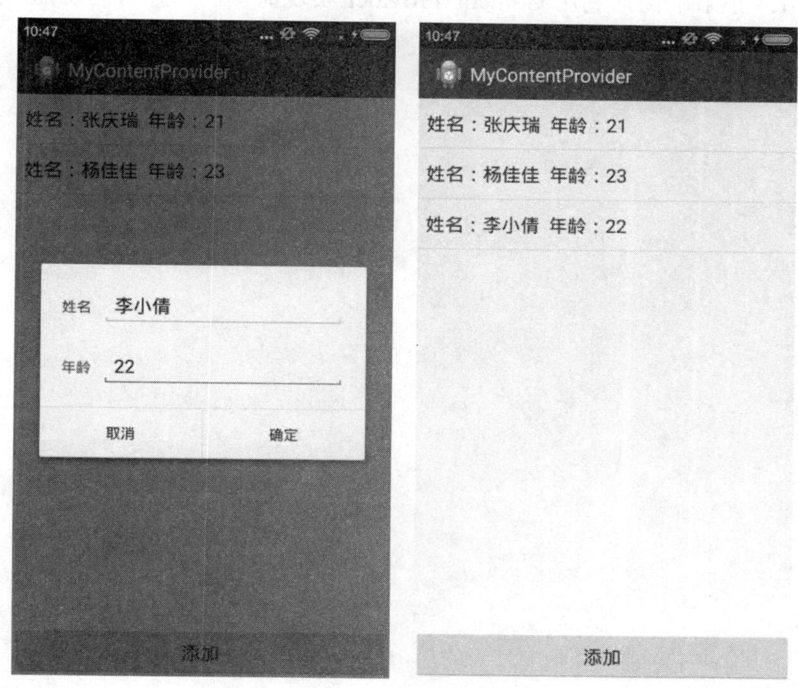

图 1-4　操作自定义的用户信息 Content Provider

本 章 小 结

(1) Content Provider 组件属于 Android 四大组件之一，用于程序间数据的共享。

(2) 使用 Content Provider 组件时，Uri 协议头和授权部分是不可或缺的。

(3) Content Provider 通常使用 SQLite 数据库来存储数据。

(4) 本质上，对 Content Provider 的操作通常是对 SQLite 数据库的操作。

本 章 练 习

(1) Android 四大组件不包括：

(A) Activity

(B) Context

(C) Service

(D) Content Provider

(2) Content Provider 协议头必须是“content://”？

(A) 是

(B) 否

(3) ____________类在 Content Provider 中担任“使用者”的角色，____________类担任“提供者”是角色。

(4) 仿照 1.4 小节示例，利用 Content Provider 实现简单的库存管理功能，并提供外界程序访问接口。

第 2 章　图形图像与动画

本章目标

- ■ 理解 Color、Paint、Path、Canvas 类的使用
- ■ 理解自定义 View 的实现
- ■ 掌握属性动画的原理及使用

在 Android 开发中，有时会发现，系统自带的控件已远远达不到或无法满足程序设计的要求，此时就需要自定义新的控件来达到预期需求。图形绘制功能是自定义控件的基础；Android 3.0 之后，新增了 Property Animation(属性动画)，功能强大，操作简单，用于弥补帧动画与补间动画的不足。本章主要介绍 Android 中的 2D 图形绘制和属性动画的运用。

2.1 图形绘制

Android 为 2D 图形操作提供了一套完整的 API，其中大部分位于“android.graphics”包中，该包中包含了四个最基本的图形绘制类，分别是：Color 类、Paint 类、Path 类和 Canvas 类。

2.1.1 Color 类

Color 类中定义了一些颜色常量和用于颜色转换的方法，Android 中的颜色分为 RGB 和 HSV 两种色彩空间，可视具体情况在两者之间转换。常用的颜色属性如表 2-1 所示。

表 2-1 Color 类中的颜色属性

属　性	描　述
BLACK	黑色
BLUE	蓝色
CYAN	青绿色
DKGRAY	灰黑色
GRAY	灰色
GREEN	绿色
LTGRAY	浅灰色
MAGENTA	红紫色
RED	红色
TRANSPARENT	透明色
WHITE	白色
YELLOW	黄色

一般在开发中使用的是 RGB 色彩空间，和 RGB 色彩空间相关的常用方法如表 2-2 所示。

表 2-2 Color 类中常用的方法

方 法 名	描　述
alpha(int color)	单独设置透明度，取值范围：0～255
red(int color)	单独设置红色值，取值范围：0～255
green(int color)	单独设置绿色值，取值范围：0～255
blue(int color)	单独设置蓝色值，取值范围：0～255
rgb(int red, int green, int blue)	同时设置红、绿、蓝色值
argb(int alpha, int red, int green, int blue)	同时设置透明度、红、绿、蓝色值
parseColor(String colorString)	将十六进制的色值转换为 RGB 模式色值，例如“#ff0000”表示红色

2.1.2　Paint 类

绘画前首先需要一支画笔，在 Android 中绘图也不例外，Paint 类就是画笔工具。在绘制前，可以设置画笔的颜色、线宽、字体、填充风格等，Paint 类也提供了一些公共的方法给开发者使用，因方法众多，常用方法如表 2-3 所示，更多方法可以查看 SDK 文档。

表 2-3　Paint 类常用方法

方 法 名	描　述
setARGB(int a,int r,int g,int b)	设置画笔的颜色，参数 a 表示透明度，r、g、b 分别表示色值，对应为：红、绿、蓝
setAlpha(int a)	设置画笔透明度
setColor(int color)	设置画笔颜色
setAntiAlias(boolean aa)	设置是否使用抗锯齿功能
setTextAlign(Paint.Align align)	设置文本对齐方式
setTextSize(float textSize)	设置字体大小
setTypeface(Typeface typeface)	设置字体风格，比如粗体、斜体等
setUnderlineText(boolean underlineText)	设置下划线效果
setStrikeThruText(boolean strikeThruText)	设置删除线效果
setStyle(Paint.Style style)	设置画笔样式，可选常量为 FILL、FILL_OR_STROKE、STROKE
setStrokeWidth(float width)	当画笔样式为 FILL_OR_STROKE 或 STROKE 时，设置画笔的线宽
setShader(Shader shader)	设置渐变效果
setShadowLayer(float radius ,float dx,float dy,int color)	设置阴影效果，radius 表示角度，dx 和 dy 表示在 x 轴和 y 轴上的距离，color 为颜色

2.1.3　Path 类

Path 类用于定义路径与线段，例如直线段、二次曲线等；也可用于设置文本路径；还可将这些线段拼接起来，形成不规则的复合图形。Path 类的常用方法如表 2-4 所示，更多方法可以查看 SDK 文档。

表 2-4　Path 类常用方法

方 法 名	描　述
addArc(RectF oval, float startAngle, float sweepAngle)	为路径添加一个多边形
addCircle(float x, float y, float radius, Path.Direction dir)	为路径添加一个圆圈
addOval(RectF oval, Path.Direction dir)	为路径添加一个椭圆形
addRect(RectF rect, Path.Direction dir)	为路径添加一个矩阵

续表

方　法　名	描　　述
addRoundRect(RectF rect, float[] radii, Path.Direction dir)	为路径添加一个圆角矩阵
isEmpty()	判断路径是否为空
transform(Matrix matrix)	应用矩阵变换
moveTo(float x, float y)	不进行绘制，只是移动画笔
lineTo(float x, float y)	定义直线，默认从(0,0)点开始
quadTo(float x1, float y1, float x2, float y2)	定义圆滑曲线，即贝塞尔曲线
rCubicTo(float x1, float y1, float x2, float y2, float x3, float y3)	定义圆滑曲线，即贝塞尔曲线(x1,y1)为控制点，(x2,y2)为控制点，(x3,y3)为结束点，比 quadTo()多了一个控制点
arcTo(RectF oval, float startAngle, float sweepAngle)	定义弧线，实际为截取所指定元的一部分，oval 为定义圆的矩形，startAngle 表示开始的角度，sweepAngle 表示旋转的角度

2.1.4 Canvas 类

正如绘画离不开纸一样，Android 中使用 Canvas 来表示画布，所有绘制都要在 Canvas 上进行。绘制之前，可以对 Canvas 画布设置一些属性，例如尺寸、颜色等。Canvas 类的主要方法如表 2-5 所示，更多方法可以查看 SDK 文档。

表 2-5　Canvas 类常用方法

方　法　名	描　　述
drawText(String text, float x, floaty, Paint paint)	绘制文本
drawRect(RectF rect, Paint paint)	绘制矩阵
drawPath(Path path, Paint paint)	绘制路径
drawBitmap(Bitmap bitmap, Rect src, Rect dst, Paint paint)	绘制图片
drawLine(float startX, float startY, float stopX, float stopY, Paintpaint)	绘制线段
drawPoint(float x, float y, Paint paint)	绘制一个点
drawOval(RectF oval, Paint paint)	绘制椭圆形
drawCircle(float cx, float cy, float radius,Paint paint)	绘制圆形
drawArc(RectF oval, float startAngle, float sweepAngle, boolean useCenter, Paint paint)	绘制弧线或区域(扇形)
clipRect(Rect rect)	裁剪矩形区域作为当前绘图区域
clipPath(Path path)	裁剪 path 范围内的区域作为当前绘图区域
rotate(float degrees)	旋转画布

2.1.5　绘制几何图形

以上小节介绍了 Android 中绘制几何图形时所需的基本工具类，本小节将介绍如何利用这些工具来绘制常见的基本图形。要想在屏幕上显示绘制的图形，首先要继承 View 类，然后重写 onDraw()方法，这样就可以像普通控件一样，加入到 Activity 中显示。

本小节将通过在画布中绘制一个“Android 机器人”示例，让读者了解绘制基本图形的方法。

1．创建自定义 View 类

新建 MyView.java 类，此类继承了 View 类，代码如下所示：

```
public class MyView extends View {

    private Paint mPaint;

    public MyView(Context context) {
        super(context);
        init();
    }

    public MyView(Context context, AttributeSet attrs) {
        super(context, attrs);
        init();
    }

    private void init() {
        // 初始化画笔并进行设置
        mPaint = new Paint();
        mPaint.setAntiAlias(true); // 消除锯齿
        // 画笔风格
        mPaint.setStyle(Paint.Style.FILL);
        mPaint.setTextSize(36); // 字体大小
    }

    @Override
    protected void onDraw(Canvas canvas) {
        super.onDraw(canvas);

        canvas.drawColor(Color.GRAY);//设置画布颜色为灰色
```

```
// 将位置移动画布指定位置
canvas.translate(canvas.getWidth() / 2 - 100, 200);

// 绘制头部
mPaint.setColor(Color.parseColor("#00d423"));
// 定义圆所在的矩阵 200*200 的正方形
RectF rectF = new RectF(0, 0, 200, 200);
// 绘制弧形实心区域，从 180 度开始，旋转 180 度结束，形成一个半圆
canvas.drawArc(rectF, 180, 180, true, mPaint);

// 绘制天线
mPaint.setStrokeWidth(10);
canvas.drawLine(0, 0, 25, 25, mPaint);
canvas.drawLine(200, 0, 175, 25, mPaint);

// 绘制眼睛
mPaint.setColor(Color.WHITE);
canvas.drawCircle(50, 50, 10, mPaint);
canvas.drawCircle(150, 50, 10, mPaint);

// 绘制身体
mPaint.setColor(Color.parseColor("#00d423"));
RectF rectF3 = new RectF(0, 110, 200, 300);
canvas.drawRoundRect(rectF3, 0, 0, mPaint);

// 绘制胳膊
canvas.drawRoundRect(new RectF(-140, 120, -10, 160),
        15, 15, mPaint);
canvas.drawRoundRect(new RectF(210, 120, 340, 160),
        15, 15, mPaint);

// 绘制腿
canvas.drawRoundRect(new RectF(20, 310, 60, 450),
        15, 15, mPaint);
canvas.drawRoundRect(new RectF(140, 310, 180, 450),
        15, 15, mPaint);

mPaint.setColor(Color.WHITE);
```

```
            mPaint.setTextAlign(Align.CENTER);
            canvas.drawText("Android", 100, 200, mPaint);
    }
}
```

上述代码中，继承了 View 类并重写了两个构造方法和 onDraw()方法。

✧ MyView(Context context)：此构造方法用于通过代码的方式创建对象。

✧ MyView(Context context, AttributeSet attrs)：此构造方法用于通过 Layout 布局文件引入的方式创建对象。

✧ onDraw(Canvas canvas)：主要回调方法，当现实 View 时调用。

2．在 activity_main.xml 布局文件中加入控件

打开 activity_main.xml 布局文件，编写代码如下：

```
<RelativeLayout xmlns:android="http://schemas.android.com/apk/res/android"
    xmlns:tools="http://schemas.android.com/tools"
android:layout_width="match_parent"
android:layout_height="match_parent" >

<com.yg.drawgraph.MyView
android:layout_width="match_parent"
android:background="#ff0000"
android:layout_height="match_parent"/>
</RelativeLayout>
```

3．运行项目

运行项目后，页面显示如图 2-1 所示。

图 2-1　图形绘制

2.2 Property Animation(属性动画)

Android3.0 之前，2D 动画分为帧动画和补间动画。帧动画是指将多张图片连续播放形成动画效果；补间动画是可以将一个视图(View)做渐变、缩放、平移、旋转操作：这两种动画都是按照预先设置好的效果执行的，只是一种视觉上的效果，细心的读者在开发中可能会发现，在对一个按钮使用补间动画做移动操作时，仅仅是视觉上做了位置的改变，按钮的实际位置并没有发生改变，并且移动后是不会触发单击事件的。

Property Animation(属性动画)是通过修改目标对象的属性来实现的，例如移动一个按钮的位置，如果通过属性动画来操作，那么这个按钮的位置就是真实地移动到指定位置。属性动画可控制的对象不只是 UI 控件，可以作用于任何对象。

与属性动画相关的常用类如表 2-6 所示。

表 2-6　属性动画常用类

类　　名	描　　述
Animator	属性动画的基类
ValueAnimator	负责初始值和结束值之间过度的计算
ObjectAnimator	对指定对象的属性进行动画操作
AnimatorSet	实现动画集合(复合动画)
AnimatorInflater	用于加载通过 XML 文件定义的属性动画

2.2.1 ValueAnimator

ValueAnimator 是属性动画中的核心类，属性动画的运行机制是通过不断改变值来实现的，而初始值和结束值之间的平滑过渡效果是由 ValueAnimator 类负责计算的。该类内部是采用一种时间循环机制来计算两个值之间的动画过渡，读者只需将初始值和结束值提供给该类，同时设置动画播放时长，该类就可以自动完成计算。除此之外，还可以设置动画播放的模式、播放次数，以及设置监听等。

ValueAnimator 类的用法并不复杂，下列代码用于实现按钮的透明度变化动画，主要代码如下：

```
ValueAnimator anim = ValueAnimator.ofFloat(1, 0, 1);
anim.setDuration(1000);
// 添加动画监听
anim.addUpdateListener(new AnimatorUpdateListener() {

    @Override
    public void onAnimationUpdate(ValueAnimator animation) {
        // 获取动画变化的属性值，并在控制台打印
```

```
            float value = (Float) animation.getAnimatedValue();
            valueBtn.setAlpha(value);
        }
});
anim.start();
```

上述代码中，首先通过 ValueAnimator 类的静态方法 ofFloat()创建对象，该方法用于设置 float 类型的值，参数为 float 类型的可变参数，通常至少为 2 个参数。ValueAnimator 会根据传入的参数两两之间进行平滑过渡的计算。接下来设置动画的播放时间为 1000 毫秒。最后为 ValueAnimator 对象添加监听器，该监听器负责监听动画过程中值的变化，当执行 start()方法时，开始执行动画。

2.2.2　ObjectAnimator

ObjectAnimator 比起 ValueAnimator 更为实用一些，因为 ValueAnimator 只是简单地对值进行一个平滑过渡动画，但实际上这种功能的使用场景并不多见。ObjectAnimator 是直接对指定对象的任意属性进行动画操作，不断地对指定对象的某个属性值进行赋值，然后根据对象属性值的变化来做动画。

下列代码用于实现通过改变按钮的“rotationY”属性值来达到旋转的动画效果，主要代码如下：

```
ObjectAnimator anim = ObjectAnimator.ofFloat(objectBtn,
        "rotationY", 0, 360, 0);
anim.setDuration(1000 * 3);
anim.start();
```

上述代码中，仅通过三行代码就能够实现按钮的旋转效果，首先通过 ObjectAnimator 的静态方法 ofFloat()创建对象，该方法参数解释如下：

- ✧ target：Object 类型，指定所要操作的任意对象。
- ✧ propertyName：String 类型，指定所要操作的对象的某个属性值。
- ✧ values：float 类型的可变参数，等同于 ValueAnimator. ofFloat()方法。

2.2.3　AnimatorSet

AnimatorSet 用于将一系列的动画进行组合，实现复合动画。也就是说，复合动画就是顺序或同时执行一系列的动画，AnimatorSet 类提供了一个 play()方法，可以向该方法中传入一个 Animator 对象(这个对象可以是 ValueAnimator 或 ObjectAnimator)，返回值是 AnimatorSet.Builder 的对象。AnimatorSet.Builder 中主要方法有：

- ✧ after(Animator anim)：将现有动画插入到该传入的动画之后执行。
- ✧ after(long delay)：将现有动画延迟指定毫秒后执行。
- ✧ before(Animator anim)：将现有动画插入到该传入的动画之前执行。

✧ with(Animator anim)：将现有动画和传入的动画同时执行。

下列代码用于实现复合动画：

```
// 设置平移动画
ObjectAnimator tranX = ObjectAnimator.ofFloat(picIv,
                "translationX", 0, 400);
ObjectAnimator tranY = ObjectAnimator.ofFloat(picIv,
                "translationY", 0,400);

// 设置缩放动画
ObjectAnimator scaleX = ObjectAnimator.ofFloat(picIv,
                "scaleX", 1.0f,0.3f);
ObjectAnimator scaleY = ObjectAnimator.ofFloat(picIv,
                "scaleY", 1.0f,0.3f);

// 设置旋转动画
ObjectAnimator rotaX = ObjectAnimator.ofFloat(picIv,
                "rotationX", 0,45, -30, 0);

// 创建动画集合对象
AnimatorSet set = new AnimatorSet();
// 添加动画并设置播放顺序
set.play(tranX).with(tranY).with(scaleX).with(scaleY).after(rotaX);
// 设置播放时间
set.setDuration(1000 * 2);
// 开始播放动画集合
set.start();
```

上述代码中，首先设置好一系列动画，然后创建一个 AnimatorSet 对象，随后通过 AnimatorSet 对象设置好动画播放的顺序并开始播放，而动画作用的对象是一个 ImageView 图片控件。

2.2.4 AnimatorInflater

除了以上几种通过编写代码方式实现的动画外，还可以通过使用 XML 文件的形式加载实现动画。加载 XML 动画资源文件就需要用到 AnimatorInflater 类，虽然使用 XML 文件编写的动画看似复杂，但是非常方便代码重用，结构清晰。

XML 属性动画布局文件通常存放到 res/animator 目录中，根节点标签有三个：

✧ <animator>：对应于 ValueAnimator。

✧ <objectAnimator>：对应于 ObjectAnimator。

✧ <set>：对应于 AnimatorSet，<set>标签之间可以互相嵌套，此标签中有个重要的属性“ordering”，取值为“sequentially”或“together”，分别表示顺序播放和同时播放。

1．<animator>

在 res/animator 目录中创建动画文件 anim_value.xml，编写代码如下：

```
<animator xmlns:android="http://schemas.android.com/apk/res/android" >
    <animator
        android:valueFrom="1.0"
        android:valueTo="0.3"
        android:duration="1000"
        android:valueType="floatType" />
</animator>
```

上述代码中，根标签为<animator>，用来定义 ValueAnimator 动画，然后在代码中引入该动画，代码如下：

```
ValueAnimator anim = (ValueAnimator) AnimatorInflater.loadAnimator(
            this, R.animator.anim_value);
// 添加动画监听
anim.addUpdateListener(new AnimatorUpdateListener() {

    @Override
    public void onAnimationUpdate(ValueAnimator animation) {
        // 获取动画变化的属性值，并在控制台打印
        float value = (Float) animation.getAnimatedValue();
        valueBtn.setAlpha(value);
    }
});
anim.start();
```

上述代码中，通过 AnimatorInflater.loadAnimator()方法引入 XML 文件定义的动画，并转换为 ValueAnimator 对象，之后的操作与使用代码创建的 ValueAnimator 动画完全相同。

2．<objectAnimator>

在 res/animator 目录中创建动画文件 anim_object.xml，编写代码如下：

```
<objectAnimator xmlns:android="http://schemas.android.com/apk/res/android" >
    <objectAnimator
        android:duration="1000"
        android:propertyName="rotationY"
        android:valueFrom="0"
```

```
            android:valueTo="360"
            android:valueType="floatType" />
</objectAnimator>
```

上述代码中，根标签为<objectAnimator>元素，用来定义 ObjectAnimator 动画，然后在代码中引入该动画，代码如下：

```
ObjectAnimator anim = (ObjectAnimator) AnimatorInflater.loadAnimator(
                this, R.animator.anim_object);
anim.setTarget(objectBtn);
anim.start();
```

上述代码中，通过 AnimatorInflater.loadAnimator()方法引入 XML 文件定义的动画，并转换为 ObjectAnimator 对象，之后的操作与使用代码创建的 ObjectAnimator 动画完全相同。

3．<set>

在 res/animator 目录中创建动画文件 anim_set.xml，编写代码如下：

```
<set xmlns:android="http://schemas.android.com/apk/res/android"
        android:ordering="sequentially" >
        <set android:ordering="sequentially" >
        <objectAnimator
                android:duration="1000"
                android:propertyName="rotationX"
                android:valueFrom="0"
                android:valueTo="45"
                android:valueType="floatType" />
        <objectAnimator
                android:duration="1000"
                android:propertyName="rotationX"
                android:valueFrom="45"
                android:valueTo="-30"
                android:valueType="floatType" />
        <objectAnimator
                android:duration="1000"
                android:propertyName="rotationX"
                android:valueFrom="-30"
                android:valueTo="0"
                android:valueType="floatType" />
</set>
<set android:ordering="together" >
```

```
    <objectAnimator
        android:duration="1000"
        android:propertyName="translationX"
        android:valueFrom="0"
        android:valueTo="400"
        android:valueType="floatType" />
    <objectAnimator
        android:duration="1000"
        android:propertyName="translationY"
        android:valueFrom="0"
        android:valueTo="400"
        android:valueType="floatType" />
    <objectAnimator
        android:duration="1000"
        android:propertyName="scaleX"
        android:valueFrom="1.0"
        android:valueTo="0.3"
        android:valueType="floatType" />
    <objectAnimator
        android:duration="1000"
        android:propertyName="scaleY"
        android:valueFrom="1.0"
        android:valueTo="0.3"
        android:valueType="floatType" />
    </set>
</set>
```

上述代码中，根标签为<set>元素，用来定义 AnimatorSet 动画，然后在代码中引入该动画，代码如下：

```
AnimatorSet set = (AnimatorSet) AnimatorInflater.loadAnimator(this,
        R.animator.anim_set);
set.setTarget(picIv);
set.start();
```

上述代码中，通过 AnimatorInflater.loadAnimator()方法引入 XML 文件定义的动画，并转换为 AnimatorSet 对象，之后的操作与使用代码创建的 AnimatorSet 复合动画完全相同。

值得注意的是，相对于 2.2.3 小节中用代码实现复合动画的方式，XML 定义动画的方式看起来比较复杂，但其具有结构清晰、方便修改、可重用性强等优点。

本 章 小 结

(1) android.graphics 包中，包含了四个最基本的图形绘制相关类，分别是 Color、Paint、Path 和 Canvas 类，Color 类主要用于管理颜色相关内容，Paint 类用于定义画笔，Path 类用于定义路径，Canvas 类用于定义一张画布。

(2) 可以通过继承 View 类来实现自定义控件的编写。

(3) 属性动画是通过修改目标对象的属性来实现的。

(4) 可以通过 AnimatorSet 类实现一系列的属性动画。

(5) 可以通过 AnimatorInflater 类加载 XML 资源文件定义的动画。

本 章 练 习

(1) 简述 Paint、Path、Canvas 三者之间的关系与作用。

(2) 简述属性动画、帧动画、补间动画之间的区别以及各自的特点。

(3) 使用属性动画，分别通过编写代码方式和 XML 文件方式实现图片的旋转。

第3章　高级网络编程

本章目标

- 理解 HttpURLConnection、HttpClient 访问网络的方法
- 了解如何上传文件到服务器
- 了解断点续传原理
- 了解如何断点续传下载文件

随着智能手机的普及，用户对网络的需求越来越丰富，例如：下载歌曲、社交聊天、远程视频等。因此，市面上大部分的智能手机应用程序也都通过网络来实现相关功能，Android 提供了一系列的网络相关 API 供开发者使用。本章主要讲解 HttpURLConnection 和 HttpClient 两个网络操作相关类，并讲解如何通过这两个类将文件上传到服务器，以及如何实现利用断点续传方式下载文件到本地。

3.1 HTTP 概述

HTTP 协议是网络通信中使用较为广泛的协议之一，随着移动互联网时代的不断完善与强大，几乎所有的移动端应用程序都离不开网络的支持，在 Android 中，针对 HTTP 的网络通信方式有两种：HttpURLConnection 和 Apache 的 HttpClient。

3.1.1 HttpURLConnection

HttpURLConnection 类继承自 URLConnection 抽象类，具有多用途、轻量级的优点，可适用于大多数应用程序。该类的实现虽然比较简单，但同时这也使得开发者能够更容易地去使用和扩展它。HttpURLConnection 常用方法如表 3-1 所示。

表 3-1 HttpURLConnection 常用方法

方 法	功 能 描 述
InputStream getInputStream()	获取当前网络请求的输入流
OutputStream getOutputStream()	获取当前网络请求的输出流
String getRequestMethod()	获取请求方法
int getResponseCode()	获取状态码，如 HTTP_OK、HTTP_UNAUTHORIZED
void setRequestMethod(String method)	设置 URL 请求的方法
void setDoInput(boolean doinput)	设置输入流，如果使用 URL 连接进行输入，则将 DoInput 标志设置为 true(默认值)；如果不使用，则为 false
void setDoOutput(boolean dooutput)	设置输出流，如果使用 URL 连接进行输出，则将 DoOutput 标志设置为 true；如果不使用，则为 false(默认值)
void setUseCaches(boolean usecaches)	设置连接是否使用任何可用的缓存
setConnectTimeout(int timeout)	设置请求连接超时毫秒数
setReadTimeout(int timeout)	设置读取数据超时毫秒数

使用 HttpURLConnection 的步骤通常如下：

(1) 创建 URL 对象，例如：

```
URL url = new URL("http://192.168.1.48:8080");
```

(2) 利用 URL 对象的 openConnection()方法返回 HttpURLConnection 对象，例如：

```
HttpURLConnection conn = (HttpURLConnection) url.openConnection();
```

(3) 设置 HTTP 请求方式，GET 方式或 POST 方式等，例如：

```
conn.setRequestMethod("GET");
```

(4) 设置请求连接超时毫秒数和读取数据超时毫秒数以及其他项，例如：

```
conn.setConnectTimeout(1000 * 10);
conn.setReadTimeout(1000 * 10);
```

(5) 调用 getInputStream()方法获得服务器返回的输入流，对输入流进行相应操作，例如：

```
InputStream in = conn.getInputStream();
```

(6) 最后调用 disconnect()方法将 HTTP 连接关掉 conn.disconnect()，例如：

```
conn.disconnect();
```

通常访问服务器的方式有两种，GET 方式或 POST 方式。

1．GET 方式访问服务器

下述示例用于实现：通过 GET 方式与服务器通信。将输入的文本内容以参数的形式上传到服务器，之后将服务器返回的文本信息显示到页面中；直接通过图片的 URL 地址从服务器下载图片并显示到页面中。

(1) 在编写 Android 客户端程序之前，需要先创建好服务器端程序，具体方法不再赘述，服务器端 Servlet 类主要代码如下：

```
@WebServlet("/DataServlet")
public class DataServlet extends HttpServlet {
	private static final long serialVersionUID = 1L;


	@Override
	protected void doGet(HttpServletRequest req, HttpServletResponse resp)
				throws ServletException, IOException {
		resp.setHeader("content-type", "text/html;charset=UTF-8");
		resp.setCharacterEncoding("UTF-8");
		req.setCharacterEncoding("UTF-8");
		String content = req.getParameter("content");
		try {
			OutputStream os = resp.getOutputStream();
			String data = "服务器消息:接收客户端(GET)消息 content="
					+ content;
			os.write(data.getBytes("UTF-8"));
			os.flush();
			os.close();
		} catch (IOException e) {
			e.printStackTrace();
```

```
            }
        }

        @Override
        protected void doPost(HttpServletRequest req,
                HttpServletResponse resp)
                        throws ServletException, IOException {
                resp.setHeader("content-type", "text/html;charset=UTF-8");
                resp.setCharacterEncoding("UTF-8");
                req.setCharacterEncoding("UTF-8");
                String name = req.getParameter("name");
                String pwd = req.getParameter("pwd");
                System.out.println(name + " - " + pwd);
                try {
                        OutputStream os = resp.getOutputStream();
                        String data = "服务器消息:接收客户端(POST)消息 name="
                                        + name + " pwd=" + pwd;
                        os.write(data.getBytes("UTF-8"));
                        os.flush();
                        os.close();

                } catch (IOException e) {
                        e.printStackTrace();
                }
        }
}
```

上述代码是服务器端的核心代码部分，接收到客户端发送的请求后，对内容进行处理，然后返回相应信息到客户端。

(2) 编写手机端程序，新建项目“ch03_HttpURLConnection”，首先创建用于 Get 方式请求的布局文件“activity_get.xml”，代码如下：

```
<LinearLayout xmlns:android="http://schemas.android.com/apk/res/android"
    xmlns:tools="http://schemas.android.com/tools"
    android:layout_width="match_parent"
    android:layout_height="match_parent"
    android:orientation="vertical" >

<RelativeLayout
    android:layout_width="match_parent"
```

```
        android:layout_height="wrap_content"
        android:orientation="horizontal" >
    <EditText
        android:id="@+id/act_get_content_et"
        android:layout_width="fill_parent"
        android:layout_height="wrap_content"
        android:layout_toLeftOf="@+id/act_get_gettext_btn"
    android:hint="input content..." />
    <Button
        android:id="@+id/act_get_gettext_btn"
        android:layout_width="wrap_content"
        android:layout_height="wrap_content"
        android:layout_alignParentRight="true"
        android:text="发送" />
    </RelativeLayout>
    <Button
        android:id="@+id/act_get_getimg_btn"
        android:layout_width="fill_parent"
        android:layout_height="wrap_content"
        android:text="获取图片" />
    <TextView
        android:id="@+id/act_get_textcontent_tv"
        android:layout_width="wrap_content"
        android:layout_height="wrap_content" />
    <ImageView
        android:id="@+id/act_get_img"
        android:layout_width="wrap_content"
        android:layout_height="wrap_content" />
</LinearLayout>
```

上述代码中，添加了一个 EditText 输入框、一个“发送”按钮与一个 TextView 文本框，用于发送数据到服务器，接收并显示返回的纯文本信息；添加了一个“获取图片”按钮和 ImageView 图片控件，用于获取并显示服务器的图片。

(3) 创建“GetActivity.java”类，首先编写基本代码，包括控件的引用和事件的添加，代码如下：

```
public class GetActivity extends Activity {

    // Handler 标识 - 显示文本信息
    protected static final int MSG_SHOW_TEXT = 1;
```

```
// Handler 标识 - 显示图片
protected static final int MSG_SHOW_IMAGE = 2;

// 输入内容的文本框
private EditText contentEt = null;
// 发送文本信息按钮
private Button sendMsgBtn = null;
// 获取图片按钮
private Button getImageBtn = null;
// 显示服务器返回的文本数据的文本框
private TextView textContentTv = null;
// 显示服务器返回的图片资源的图片控件
private ImageView imageIv = null;

@Override
protected void onCreate(Bundle savedInstanceState) {
        super.onCreate(savedInstanceState);
        setContentView(R.layout.activity_get);

        contentEt = (EditText) findViewById(R.id.act_get_content_et);
        sendMsgBtn = (Button) findViewById(R.id.act_get_gettext_btn);
        getImageBtn = (Button) findViewById(R.id.act_get_getimg_btn);
        textContentTv =
                (TextView) findViewById(R.id.act_get_textcontent_tv);
        imageIv = (ImageView) findViewById(R.id.act_get_img);

        // 添加按钮点击事件
        sendMsgBtn.setOnClickListener(onClickListener);
        getImageBtn.setOnClickListener(onClickListener);
}
private OnClickListener onClickListener = new OnClickListener() {
        @Override
        public void onClick(View view) {
                if (view == sendMsgBtn) {
                        String content = contentEt.getText().toString();
                        try {
                                // 将输入的内容以 UTF-8 编码方式进行 URL 编码
                                content = URLEncoder.encode(content, "UTF-8");
                        } catch (UnsupportedEncodingException e) {
```

```
                        e.printStackTrace();
                    }
                    String path = "http://192.168.1.48:8080/"
                        + "HttpService/DataServlet?content="+ content;
                    getServiceTextContent(path);
                } else if (view == getImageBtn) {
                    String path = "http://192.168.1.48:8080/"
                        + "HttpService/pic.png";
                    getServiceImage(path);
                }
            }
        };
}
```

当点击“发送”按钮时，调用 sendContentToService()方法，将输入的文本内容发送到服务器；当点击“获取图片”按钮时，调用 getServiceImage()方法从服务器下载图片并显示。sendContentToService()方法和 getServiceImage()方法代码如下：

```
/**
 * 发送文本数据到服务器
 */
private void sendContentToService(final String path) {
    new Thread(new Runnable() {

        @Override
        public void run() {
            // 将 byte 数组转换为 String，编码方式采用 UTF-8
            String reqStr = "";
            try {
                reqStr = new String(getDataForGet(path), "UTF-8");
            } catch (UnsupportedEncodingException e) {
                e.printStackTrace();
            }
            Message msg = new Message();
            msg.what = MSG_SHOW_TEXT;
            msg.obj = reqStr;
            handler.sendMessage(msg);
        }
    }).start();
}
```

```
/**
 *获取服务器图片
 */
private void getServiceImage(final String path) {
        new Thread(new Runnable() {

                @Override
                public void run() {
                        byte[] data = getDataForGet(path);
                        Bitmap bitmap = BitmapFactory.decodeByteArray(data, 0,
                                data.length);
                        Message msg = new Message();
                        msg.what = MSG_SHOW_IMAGE;
                        msg.obj = bitmap;
                        handler.sendMessage(msg);
                }
        }).start();
}
```

上述两个方法中，都有一个共同的方法是 getDataForGet()方法，该方法用于访问服务器，将服务器返回的数据以 byte[]数组的形式返回给调用者，调用者获取到该数据后，根据实际需求进行操作。getDataForGet()方法代码如下：

```
/**
 * GET 方式访问服务器获取数据
 */
private byte[] getDataForGet(String path) {
        try {
                // 创建 URL
                URL url = new URL(path);
                // 创建 HttpURLConnection 对象
                HttpURLConnection conn = (HttpURLConnection)url.openConnection();
                // 设置为 GET 方式请求
                conn.setRequestMethod("GET");
                // 设置连接超时
                conn.setConnectTimeout(1000 * 10);
                // 判断响应码，如果是 HttpStatus.SC_OK(200)，则说明请求成功
                if (conn.getResponseCode() == HttpStatus.SC_OK) {
                        // 获取输入流
                        InputStream in = conn.getInputStream();
```

```
                    // 将输入流转换为 byte 数组
                    byte[] data = readData(in);
                    conn.disconnect();
                    return data;
                }
            } catch (IOException e) {
                e.printStackTrace();
            }
            return null;
}

/**
 * 从流中读取数据
 */
publicbyte[] readData(InputStream inStream) throws IOException {
        ByteArrayOutputStream outStream = new ByteArrayOutputStream();
        byte[] buffer = new byte[1024];
        int len = 0;
        while ((len = inStream.read(buffer)) != -1) {
                outStream.write(buffer, 0, len);
        }
        inStream.close();
        return outStream.toByteArray();
}
```

通过上述代码，getDataForGet()中获取输入流后，通过 readData()方法将该输入流解析为 byte[]数据并返回。

最后，需要将服务器返回的数据更新到界面上显示，由于 Android 程序中，网络操作部分必须放到子线程中执行，子线程不能直接更新 UI，因此可以通过 Handler 更新 UI。创建 Handler 对象，代码如下：

```
Handler handler = new Handler() {
        public void handleMessage(Message msg) {
                switch (msg.what) {
                case MSG_SHOW_TEXT:
                        String contentStr = (String) msg.obj;
                        textContentTv.setText(contentStr);
                        break;
                case MSG_SHOW_IMAGE:
```

```
                    Bitmap bmp = (Bitmap) msg.obj;
                    imageIv.setImageBitmap(bmp);
                    break;
            }
        };
};
```

至此，使用 HttpURLConnection 的 GET 方式与服务器通信的功能已完成。

(4) 在运行之前，需要在 AndroidManifest.xml 中添加网络访问权限，代码如下：

```
<uses-permission android:name="android.permission.INTERNET" />
```

(5) 运行程序，效果如图 3-1 所示。

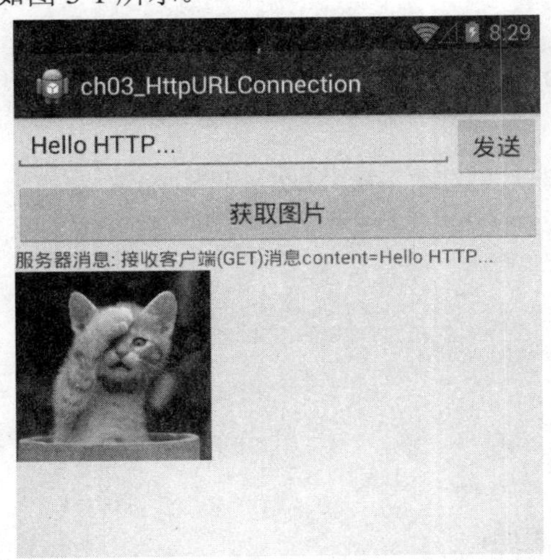

图 3-1　HttpURLConnection GET 方式请求服务器

2. POST 方式访问服务器

当开发者希望上传文件等大量数据到服务器，使用 GET 方式是无法完成的，此时就需要使用 POST 方式来完成相应操作。相对于 GET 方式，POST 方式上传数据更加安全。

下述示例用于实现：点击“提交”按钮，将用户输入的“用户名”和“密码”以 POST 方式上传到服务器，将服务器返回的信息显示到界面上。

(1) 该示例依然在“ch03_HttpURLConnection”项目中实现，首先创建布局文件“activity_post.xml”，代码如下：

```
<LinearLayout xmlns:android="http://schemas.android.com/apk/res/android"
    xmlns:tools="http://schemas.android.com/tools"
    android:layout_width="match_parent"
    android:layout_height="match_parent"
    android:orientation="vertical" >

<EditText
```

```
    android:id="@+id/act_post_name_et"
    android:layout_width="fill_parent"
    android:layout_height="wrap_content"
    android:hint="input name..." />
<EditText
    android:id="@+id/act_post_pwd_et"
    android:layout_width="fill_parent"
    android:layout_height="wrap_content"
    android:hint="input password..."
    android:inputType="textPassword" />
<Button
    android:id="@+id/act_post_send_btn"
    android:layout_width="fill_parent"
    android:layout_height="wrap_content"
    android:text="提交" />
<TextView
    android:id="@+id/act_post_req_et"
    android:layout_width="wrap_content"
    android:layout_height="wrap_content"/>
</LinearLayout>
```

上述代码中，添加了两个 EditText 输入框，分别用于输入“用户名”和“密码”；添加了一个“提交”按钮，用于将数据以 POST 方式提交到服务器；最后添加了一个 TextView 文本框，用于显示服务器返回的数据。

(2) 创建“PostActivity.java”类，该类用户通过 POST 方式请求服务器。首先编写基本代码，包括控件的引用以及事件等，代码如下：

```
public class CopyOfPostActivity extends Activity {
        // Handler 标识 - 显示文本信息
        protected static final int MSG_SHOW_TEXT = 1;
        private Button sendBtn = null;
        private EditText nameEt = null;
        private EditText pwdEt = null;
        private TextView reqContentTv = null;

        @Override
        protected void onCreate(Bundle savedInstanceState) {
                super.onCreate(savedInstanceState);
                setContentView(R.layout.activity_post);

                sendBtn = (Button) findViewById(R.id.act_post_send_btn);
```

```
        nameEt = (EditText) findViewById(R.id.act_post_name_et);
        pwdEt = (EditText) findViewById(R.id.act_post_pwd_et);
        reqContentTv = (TextView) findViewById(R.id.act_post_req_et);
        sendBtn.setOnClickListener(new OnClickListener() {

            @Override
            public void onClick(View view) {
                String path =
                "http://192.168.1.48:8080/HttpService/DataServlet";
                String name = nameEt.getText().toString();
                String pwd = pwdEt.getText().toString();
                getDataForPost(path, name, pwd);
            }
        });
    }
}
```

当点击“提交”按钮时，调用 getDataForPost()方法，将户名和密码提交到服务器，该方法代码如下：

```
/**
 * POST 方式访问服务器
 */
public void getDataForPost(final String path, final String name,
        final String pwd) {
    new Thread(new Runnable()
    {
        @Override
        public void run() {
            try {
                URL url = new URL(path);
                HttpURLConnection conn = (HttpURLConnection) url.openConnection();
                // 设置为 POST 方式请求
                conn.setRequestMethod("POST");
                // 设置连接超时
                conn.setConnectTimeout(1000 * 10);
                // 设置读取超时
                conn.setReadTimeout(1000 * 10);
                // 设置允许输入、输出
                conn.setDoOutput(true);
                conn.setDoInput(true);
```

```
                    // 不适用缓存
                    conn.setUseCaches(false);
                    // 参数拼接
                    String data = "name="+URLEncoder.encode(name, "UTF-8")
                            + "&pwd=" + URLEncoder.encode(pwd, "UTF-8");

                    // 获取输出流
                    OutputStream out = conn.getOutputStream();
                    // 写入参数
                    out.write(data.getBytes());
                    out.flush();
                    // 判断响应码，如果是 HttpStatus.SC_OK(200)，则说明请求成功
                    if (conn.getResponseCode() == HttpStatus.SC_OK)
                    {
                        // 获取响应的输入流对象
                        InputStream is = conn.getInputStream();
                        // 返回字符串
                        String reqStr = new String(readData(is));
                        Message msg = new Message();
                        msg.what = MSG_SHOW_TEXT;
                        msg.obj = reqStr;
                        handler.sendMessage(msg);
                    }
                } catch (Exception e)
                {
                    e.printStackTrace();
                }
            }
        }).start();
}
```

该方法为核心方法，当获取到服务器返回的数据时，依然通过调用 readData()方法将输入流解析为 byte[]数据返回，readData()方法不再赘述。

(3) 添加用于更新界面的 Handler 对象，代码如下：

```
Handler handler = new Handler() {
    public void handleMessage(Message msg) {
        switch (msg.what) {
        case MSG_SHOW_TEXT:
            String contentStr = (String) msg.obj;
            reqContentTv.setText(contentStr);
```

```
                    break;
                }
        };
};
```

至此，使用 HttpURLConnection POST 方式与服务器通信的功能已完成。运行效果如图 3-2 所示。

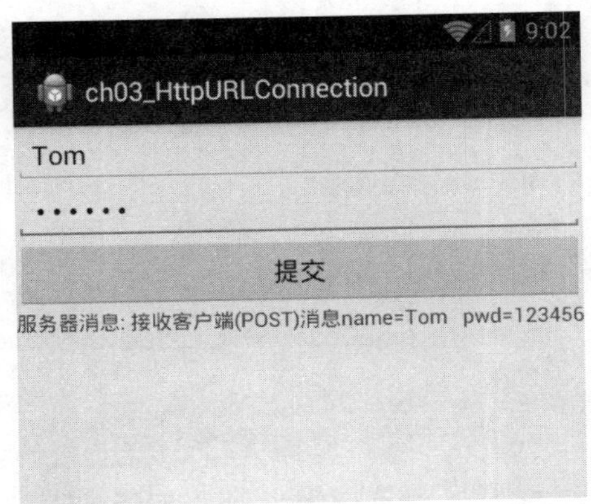

图 3-2　HttpURLConnection POST 方式请求服务器

3.1.2　HttpClient

Apache 提供了 HTTP 客户端组件 HttpClient，它对 java.net 包中的类进行封装和抽象，更适合在 Android 上开发网络应用，使得针对 HTTP 编程更加方便、高效。HttpClient 本身不是一个浏览器，而是一个客户端的 HTTP 传输库，用于接收和发送 HTTP 消息。HttpClient 主要用于执行 HTTP 方法，执行一个 HTTP 方法会涉及一个或几个 HTTP 请求或响应的交互，根据请求方法的不同会用到 HttpGet 和 HttpPost 两个对象，而 HttpClient 负责将该项请求转送到目标服务器并返回一个相应的响应对象。因此，HttpClient API 的主要部分是定义上述功能的 HttpClient 接口。HttpClient 接口代表了最基本的 HTTP 请求执行规约。HttpClient 没有在请求执行的过程上强加任何限制或特定的具体细节，不关心连接管理细节，状态管理细节，认证和重定向处理个别的实现。这使得使用额外的功能实现接口变得容易。

通常使用 HttpClient 的子类 DefaultHttpClient 进行操作，DefaultHttpClient 是 HttpClient 的默认实现类，用来负责处理 HTTP 协议的某一方面功能，如重定向或认证处理、关于保持连接和保活时间的决策。这使得用户可以选择性地使用特定的应用替换默认的功能。

使用 HttpClient 的步骤通常如下：

(1) 创建 HttpClient 对象，例如：

```
HttpClient httpClient = new DefaultHttpClient();
```

(2) 设置请求连接超时毫秒数和读取数据超时毫秒数及其他项，例如：

```
HttpParams params = httpClient.getParams();
```

```
params.setParameter(CoreConnectionPNames.CONNECTION_TIMEOUT, 1000 * 10);
params.setParameter(CoreConnectionPNames.SO_TIMEOUT, 1000 * 10);
```

(3) 如果使用 GET 方式创建 HttpGet 对象，则使用 POST 方式创建 HttpPost 对象。创建 HttpGet 对象，例如：

```
HttpGet httpGet = new HttpGet("http://192.168.1.48:8080/"
                                    +"HttpService/DataServlet");
```

创建 HttpPost 对象，例如：

```
HttpPost httpPost = new HttpPost("http://192.168.1.48:8080/"
                                    +"HttpService/DataServlet");
```

(4) 调用 HttpClient 对象的 execute()方法访问服务器，该方法需传入 HttpGet 对象或 HttpPost 对象，返回值为 HttpResponse 对象，例如：

```
HttpResponse httpResponse = httpClient.execute(httpGet);
```

(5) 通过调用返回的 HttpResponse 对象的 getEntity()方法获取 HttpEntity 对象，该对象包装了服务器的相应内容，例如：

```
HttpEntity entity = httpResponse.getEntity();
String reqStr = EntityUtils.toString(entity, "utf-8");
```

访问服务器的方式通常有两种，GET 方式和 POST 方式。

1. GET 方式访问服务器

下述示例用于实现：通过 HttpClient 的 GET 方式实现与 3.1.1 小节中使用 HttpURLConnection 的 GET 方式访问服务器相同的功能。

(1) 创建项目“ch03_HttpClient”，将“ch03_HttpURLConnection”项目中的“activity_get.xml”布局文件复制到该项目中。创建“GetActivity.java”类，该类通过 GET 方式与服务器交互，首先编写基本代码如下：

```
public class GetActivity extends Activity {

    // 属性、变量声明部分同 ch03_HttpURLConnection 项目中
    // GetActivity.java 部分

    @Override
    protected void onCreate(Bundle savedInstanceState) {
            super.onCreate(savedInstanceState);
            setContentView(R.layout.activity_get);

            //同 ch03_HttpURLConnection 项目中 GetActivity.java 部分
    }

    private OnClickListener onClickListener = new OnClickListener() {
```

```
        @Override
        public void onClick(View view) {
            if (view == sendMsgBtn) {
                String content = contentEt.getText().toString();
                String path = "http://192.168.1.48:8080/" + "HttpService/DataServlet";
                sendContentToService(path, content);
            } else if (view == getImageBtn) {
                String path = "http://192.168.1.48:8080/" + "HttpService/pic.png";
                getServiceImage(path);
            }
        }
    };
}
```

当点击“发送”按钮时，调用 sendContentToService()方法，将用户输入的文本信息发送到服务器，该方法代码如下：

```
/**
 * 发送文本数据到服务器
 */
private void sendContentToService(final String path,final String content){
    new Thread(new Runnable() {
        @Override
        public void run() {
            try {
                // 创建参数列表
                List<BasicNameValuePair> params =
                        new LinkedList<BasicNameValuePair>();
                params.add(new BasicNameValuePair("content", content));
                // 对参数进行 URL 编码
                String param = URLEncodedUtils.format(params, "UTF-8");
                // 创建 HttpClient 对象
                HttpClient httpClient = new DefaultHttpClient();
                // 设置连接超时
                HttpParams httpParams = httpClient.getParams();
                httpParams.setParameter(
                        CoreConnectionPNames.CONNECTION_TIMEOUT, 1000 * 10);
                httpParams.setParameter(
                        CoreConnectionPNames.SO_TIMEOUT,1000 * 10);
                // 创建 HttpGet 对象，同时拼接请求地址加参数
                HttpGet httpGet = new HttpGet(path + "?" + param);
```

```
                        // 执行网络访问，获取 HttpResponse 对象
                        HttpResponse httpResponse =  httpClient.execute(httpGet);
                        String reqStr = "连接服务器失败";
                        // 判断响应码，HttpStatus.SC_OK(200)表示响应成功
                        if (httpResponse.getStatusLine().getStatusCode() ==
                                HttpStatus.SC_OK) {
                                // 获取响应结果的实体对象
                                HttpEntity entity = httpResponse.getEntity();
                                // 讲响应结果转换为字符串
                                reqStr = EntityUtils.toString(entity, "utf-8");
                        }
                        Message msg = new Message();
                        msg.what = MSG_SHOW_TEXT;
                        msg.obj = reqStr;
                        handler.sendMessage(msg);
                } catch (IOException e) {
                        e.printStackTrace();
                }
            }
        }).start();
}
```

当点击“获取图片”按钮时，调用 getServiceImage()方法，获取服务器图片，该方法代码如下：

```
/**
 * 获取服务器图片
 */
private void getServiceImage(final String path) {
        new Thread(new Runnable() {
                @Override
                public void run() {
                        try {
                                Bitmap bitmap = null;
                                // 创建 HttpClient 对象
                                HttpClient httpClient = new DefaultHttpClient();
                                // 设置连接超时
                                HttpParams httpParams = httpClient.getParams();
                                httpParams.setParameter(
                                        CoreConnectionPNames.CONNECTION_TIMEOUT,1000*10);
                                httpParams.setParameter(CoreConnectionPNames.SO_TIMEOUT,1000 * 10);
```

```
                    // 创建 HttpGet 对象
                    HttpGet get = new HttpGet(path);
                    // 执行网络访问，获取 HttpResponse 对象
                    HttpResponse response = httpClient.execute(get);
                    // 判断响应码，HttpStatus.SC_OK(200)表示响应成功
                    if (response.getStatusLine().getStatusCode() == HttpStatus.SC_OK) {
                        // 从响应结果的实体对象中获取输入流
                        InputStream is = response.getEntity().getContent();
                        // 通过输入流创建 Bitmap 对象
                        bitmap = BitmapFactory.decodeStream(is);
                        is.close();
                    }
                    Message msg = new Message();
                    msg.what = MSG_SHOW_IMAGE;
                    msg.obj = bitmap;
                    handler.sendMessage(msg);
                } catch (IOException e) {
                    e.printStackTrace();
                }
            }
        }).start();
}
```

最后，将服务器返回的信息，通过 Handler 更新到界面中，Handler 对象代码如下：

```
Handler handler = new Handler() {
    public void handleMessage(Message msg) {
        switch (msg.what) {
        case MSG_SHOW_TEXT:
            String contentStr = (String) msg.obj;
            textContentTv.setText(contentStr);
            break;
        case MSG_SHOW_IMAGE:
            Bitmap bmp = (Bitmap) msg.obj;
            imageIv.setImageBitmap(bmp);
            break;
        }
    };
};
```

至此，使用 HttpClient 的 GET 方式与服务器通信的功能已完成，运行结果同图 3-1。

2. POST 方式访问服务器

下述示例用于实现：通过 HttpClient 的 POST 方式实现与 3.1.1 小节中使用 HttpURLConnection 的 POST 方式访问服务器相同的功能。

(1) 在“ch03_HttpClient”项目中，将“ch03_HttpURLConnection”项目中的“activity_post.xml”布局文件复制到该项目中。创建“PostActivity.java”类，该类通过 POST 方式与服务器交互，首先编写基本代码如下：

```
public class PostActivity extends Activity {
    // 属性、变量声明部分同 ch03_HttpURLConnection 项目中
    // PostActivity.java 部分

    @Override
    protected void onCreate(Bundle savedInstanceState) {
        super.onCreate(savedInstanceState);
        setContentView(R.layout.activity_post);

        //同 ch03_HttpURLConnection 项目中 PostActivity.java 部分
    }
}
```

(2) 在该类中添加 getDataForPost()方法，用于将用户输入的“用户名”和“密码”以 POST 方式发送到服务器，代码如下：

```
/**
 * POST 方式访问服务器
 */
public void getDataForPost(final String path, final String name,
        final String pwd) {
    new Thread(new Runnable() {
        @Override
        public void run() {
            try {
                // 创建参数列表
                List<BasicNameValuePair> params = new LinkedList<BasicNameValuePair>();
                params.add(new BasicNameValuePair("name", name));
                params.add(new BasicNameValuePair("pwd", pwd));
                // 创建 HttpClient 对象
                HttpClient httpClient = new DefaultHttpClient();
                // 设置连接超时
                HttpParams httpParams = httpClient.getParams();
                httpParams.setParameter(
                        CoreConnectionPNames.CONNECTION_TIMEOUT,
```

```
                        1000 * 10);
                httpParams.setParameter(CoreConnectionPNames.SO_TIMEOUT,1000 * 10);
                // 创建 HttpPost 对象
                HttpPost httpPost = new HttpPost(path);
                // 对参数进行 URL 编码
                UrlEncodedFormEntity paramEntity =
                        new UrlEncodedFormEntity(params, "UTF-8");
                // 将编码后的参数对象添加到 HttpPost 中
                httpPost.setEntity(paramEntity);
                // 执行网络访问，获取 HttpResponse 对象
                HttpResponse httpResponse =
                        httpClient.execute(httpPost);
                String reqStr = "连接服务器失败";
                // 判断响应码，HttpStatus.SC_OK(200)表示响应成功
                if (httpResponse.getStatusLine().getStatusCode()
                         == HttpStatus.SC_OK) {
                        // 获取响应结果的实体对象
                        HttpEntity entity = httpResponse.getEntity();
                        // 讲响应结果转换为字符串
                        reqStr = EntityUtils.toString(entity, "utf-8");
                }
                Message msg = new Message();
                msg.what = MSG_SHOW_TEXT;
                msg.obj = reqStr;
                handler.sendMessage(msg);
            } catch (IOException e) {
                e.printStackTrace();
            }
        }
    }).start();
}
```

(3) 创建 Hanlder 对象，用于讲服务器数据显示到界面，代码如下：

```
Handler handler = new Handler() {
    public void handleMessage(Message msg) {
        switch (msg.what) {
        case MSG_SHOW_TEXT:
            String contentStr = (String) msg.obj;
            reqContentTv.setText(contentStr);
```

```
                break;
            }
        };
};
```

至此，使用 HttpClient 的 POST 方式与服务器通信的功能已完成。运行结果同图 3-2。

3.2 上传文件到服务器

在 Android 开发中，不可避免地会遇到文件上传的问题，而普通的网络请求方式无法实现，遇到这种情况，目前通常是通过导入第三方 jar 包来实现文件。本小节将采用原生代码的方式来实现此功能，即通过 HttpConnection 模拟网页表单的方式来上传文件。

下述示例用于实现：将内存卡中的文件“readme.txt”上传到服务器，并使用 Toast 显示服务器上传结果。

(1) 首先实现服务器端功能，主要 Servlet 代码如下：

```
@WebServlet("/UploadServlet")
public class UploadServlet extends HttpServlet {
    private static final long serialVersionUID = 1L;
    private String path;

    public void doGet(HttpServletRequest request,HttpServletResponse response)
                throws ServletException, IOException {
        this.doPost(request, response);
    }

    protected void doPost(HttpServletRequest request,HttpServletResponse
            response) throws ServletException, IOException {
            response.setContentType("text/html;charset=utf-8");
            request.setCharacterEncoding("utf-8");
            response.setCharacterEncoding("utf-8");
            PrintWriter out = response.getWriter();
            DiskFileItemFactory factory = new DiskFileItemFactory();
            // 文件保存路径
            String upload = "d:\\";
            // 获取系统默认临时文件路径
            String temp = System.getProperty("java.io.tmpdir");
            // 设置缓冲区大小(3M)
            factory.setSizeThreshold(1024 * 1024 * 3);
            // 设置临时文件夹为 temp
```

```
        factory.setRepository(new File(temp));
        ServletFileUpload servletFileUpload = new ServletFileUpload(factory);
        try {
            List<FileItem> list = (List<FileItem>) servletFileUpload
                    .parseRequest(new ServletRequestContext(request));

            for (FileItem item : list) {
                String name = item.getFieldName();
                InputStream is = item.getInputStream();

                if (name.contains("file")) {
                    try {
                        path = upload + item.getName();
                        BuildFileForInputStream(is, path);
                        System.out.println("文件保存位置： " + path);
                        break;
                    } catch (Exception e) {
                        e.printStackTrace();
                    }
                }
            }
            out.write("服务器：文件上传成功");
        } catch (FileUploadException e) {
            e.printStackTrace();
            out.write("服务器：文件上传失败");
        }

        out.flush();
        out.close();
    }

    /**
     * 通过输入流创建文件
     *
     * @param is
     * @param savePath
     * @throws Exception
     */
    public static void BuildFileForInputStream(InputStream inputSteam,
```

```
                    String savePath) throws Exception {
            BufferedInputStream fis = new BufferedInputStream(inputSteam);
            FileOutputStream fos = new FileOutputStream(new File(savePath));
            int len;
            while ((len = fis.read()) != -1) {
                    fos.write(len);
            }
            fos.flush();
            fos.close();
            fis.close();
            inputSteam.close();
    }
}
```

(2) 编写手机端程序，创建项目“ch03_uploadFile”，修改“activity_main.xml”布局文件，代码如下：

```
<RelativeLayout xmlns:android="http://schemas.android.com/apk/res/android"
    xmlns:tools="http://schemas.android.com/tools"
    android:layout_width="match_parent"
    android:layout_height="match_parent">
<Button
    android:id="@+id/act_main_btn"
    android:layout_width="fill_parent"
    android:layout_height="wrap_content"
    android:text="上传文件" />
</RelativeLayout>
```

在该布局中，仅添加了一个“上传文件”按钮。

(3) 编写 MainActivity.java 类，首先编写基本代码如下：

```
public class MainActivity extends Activity {

    private static final int TAG_SHOW_MESSAGE = 1;

    @Override
    public void onCreate(Bundle savedInstanceState) {
            super.onCreate(savedInstanceState);
            setContentView(R.layout.activity_main);

            Button btn = (Button) findViewById(R.id.act_main_btn);
            btn.setOnClickListener(new OnClickListener() {
                    @Override
```

```
                public void onClick(View view) {
                    initUploadFile();
                }
            });
    }
}
```

点击“上传文件”按钮时，调用 initUploadFile()方法，准备要上传到服务器的文件，并执行上传操作，该方法代码如下：

```
/**
 * 准备将文件上传到服务器
 */
protected void initUploadFile() {
    // 获取本地文件
    final File file = new File(Environment.getExternalStorageDirectory()
            + "/readme.txt");
    new Thread(new Runnable()
    {
        @Override
        public void run() {
            // 开始上传文件
            String request = uploadFile(file,
            http://192.168.1.48:8080/+ "HttpService/UploadServlet");
            // 将服务器返回的消息显示到界面
            Message msg = new Message();
            msg.what = TAG_SHOW_MESSAGE;
            msg.obj = request;
            handler.sendMessage(msg);
        }
    }).start();
}
Handler handler = new Handler() {
    public void dispatchMessage(Message msg) {
        if (msg.what == TAG_SHOW_MESSAGE)
        {
            Toast.makeText(getApplicationContext(), (String) msg.obj,
                    Toast.LENGTH_LONG).show();
        }
    };
};
```

该方法中调用 uploadFile()方法进行文件上传工作，并接收上传结果。uploadFile()方法代码如下：

```
/**
 * 上传文件到服务器
 */
public String uploadFile(File file, String urlStr) {
	String result = "";
	String BOUNDARY = UUID.randomUUID().toString();
	String PREFIX = "--", LINE_END = "\r\n";
	String CONTENT_TYPE = "multipart/form-data";

	try {
		URL url = new URL(urlStr);
		HttpURLConnection conn =
			(HttpURLConnection) url.openConnection();
		conn.setReadTimeout(1000 * 10);
		conn.setConnectTimeout(1000 * 10);
		conn.setDoInput(true); // 允许输入流
		conn.setDoOutput(true); // 允许输出流
		conn.setUseCaches(false); // 禁用缓存
		conn.setRequestMethod("POST");
		// 设置编码方式为 utf-8
		conn.setRequestProperty("Charset", "utf-8");
		conn.setRequestProperty("connection", "keep-alive");
		conn.setRequestProperty("Content-Type", CONTENT_TYPE +
			";boundary="+ BOUNDARY);
		conn.connect();

		if (file != null) {
			// 模拟表单
			DataOutputStream dos = new DataOutputStream(
					conn.getOutputStream());
			StringBuffer sb = new StringBuffer();
			sb.append(PREFIX);
			sb.append(BOUNDARY);
			sb.append(LINE_END);
			sb.append("Content-Disposition: form-data;name=\"file\";
				filename=\""+ file.getName() + "\"" + LINE_END);
			sb.append("Content-Type: application/octet-stream;
```

```
                    charset=utf-8"+ LINE_END);
            sb.append(LINE_END);
            dos.write(sb.toString().getBytes());
            InputStream is = new FileInputStream(file);
            byte[] bytes = new byte[1024];
            int len = 0;
            while ((len = is.read(bytes)) != -1) {
                dos.write(bytes, 0, len);
            }
            is.close();
            dos.write(LINE_END.getBytes());
            byte[] end_data = (PREFIX + BOUNDARY + PREFIX + LINE_END).getBytes();
            dos.write(end_data);
            dos.flush();
            // 判断响应码，HttpStatus.SC_OK 表示响应成功
            int res = conn.getResponseCode();
            if (res == HttpStatus.SC_OK) {
                // 获取输入流
                InputStream in = conn.getInputStream();
                // 将输入流转换为 byte 数组
                byte[] data = readData(in);
                result = new String(data, "utf-8");
            }
        }
    } catch (IOException e) {
        e.printStackTrace();
    }
    return result;
}

/**
 * 从流中读取数据
 */
public byte[] readData(InputStream inStream) throws IOException {
    ByteArrayOutputStream outStream = new ByteArrayOutputStream();
    byte[] buffer = new byte[1024];
    int len = 0;
    while ((len = inStream.read(buffer)) != -1) {
        outStream.write(buffer, 0, len);
```

```
        }
        inStream.close();
        return outStream.toByteArray();
}
```

uploadFile()方法作为程序的核心方法，上传文件功能通过模拟表单的方式实现。

(4) 在“AndroidManifest.xml”配置文件中，添加网络访问权限和文件读取权限，代码如下：

```
<uses-permission android:name=
"android.permission.READ_EXTERNAL_STORAGE" />
<uses-permission android:name="android.permission.INTERNET" />
```

(5) 启动服务器，运行本客户端示例，当单击“上传文件”按钮时，将本地文件上传到服务器，并将服务器返回的信息显示到屏幕。上传的文件将会保存到服务器的 D 盘根目录下，读者可以打开文件进行验证。

注 意　简单起见，本示例没有加入选择文件功能，“readme.txt”文件需要事先拷贝到手机内存卡根目录下。

3.3　断点续传下载文件

目前移动设备主要通过 Wi-Fi 和手机 SIM 卡两种方式连接网络，而使用手机 SIM 卡上网需要耗费大量的流量，以至于用户话费剧增。为了能够尽量不浪费宝贵的流量，同时也为了节省更多的时间和服务器的压力，Android 网络相关 API 提供了断点续传下载或上传文件技术。为了简单易懂起见，本小节只讲述单线程断点续传下载文件，读者可在此基础上，实现多线程下载文件的功能。

3.3.1　断点续传的流程及原理

单线程断点续传下载文件的原理比较简单，在下载文件过程中，可能由于各种原因导致下载中断，以至于无法完整地下载文件，当再次下载时，能够从之前的下载位置继续下载文件，直至完整下载文件下完。需要注意的是，要使用断点续传功能，必须要有服务器的支持。

由于断点续传不同于普通下载方式，它需要时刻保存当前的下载进度信息，以便中断下载后，继续下载剩余数据，所以用于保存这些信息最合适的介质是数据库，本小节的示例中，也将使用数据库来存储下载信息。

断点续传的流程如图 3-3 所示，当用户执行下载命令后，首先检查本地数据库是否存在对应的记录，如果不存在记录，则将创建一条新的记录和本地文件，然后开始下载；如果已经存在记录，则将先解析记录信息，然后继续下载，当文件完全下载完毕，删除数据库中对应的记录。

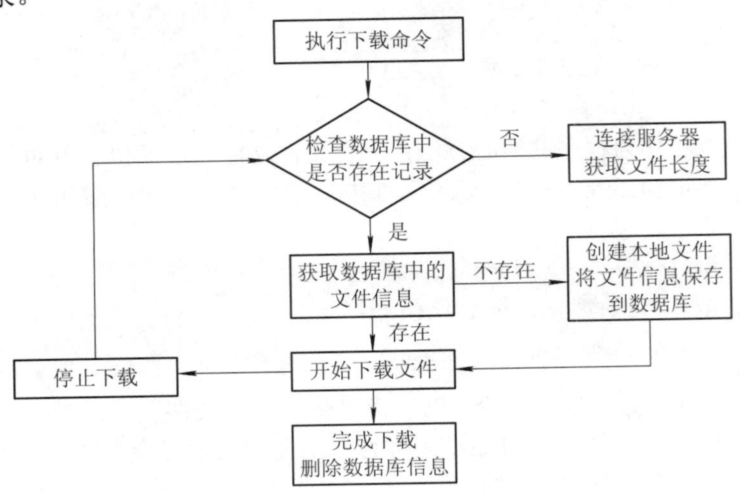

图 3-3　断点续传的流程图

3.3.2　断点续传的实现

下述示例用于实现：使用单线程断点续传方式下载服务器文件“music.mp3”。

1. 服务器端的实现

在实现客户端断点续传功能之前，需要有服务器的支持，因此首先实现服务器端代码，简单起见，本示例中将固定下载文件为“F:/music.mp3”，编写服务器端代码如下：

```
@WebServlet("/DownloadServlet")
public class DownloadServlet extends HttpServlet {
	private static final long serialVersionUID = 1L;

	@Override
	protected void doGet(HttpServletRequest req, HttpServletResponse resp)
			throws ServletException, IOException {
		this.doPost(req, resp);
	}

	@Override
	protected void doPost(HttpServletRequest request,
			HttpServletResponse response) {

		String path = "F:/music.mp3";
		File downloadFile = new File(path);
		if (!downloadFile.exists()) {
			System.out.println("文件不存在");
```

```
                return;
        }
        // 文件总大小
        long totalLength = downloadFile.length();
        // 已下载文件大小
        long compeleteLength = 0;
        int rangeSwitch = 0;
        long toLength = 0;
        long contentLength = 0;
        String rangeBytes = "";
        RandomAccessFile raf = null;
        OutputStream os = null;
        OutputStream bufferOs = null;
        byte b[] = new byte[1024 * 1024];
        // 获取请求头重的“Range”信息，得到下载的文件块的开始位置
        if (request.getHeader("Range") != null) {
                response.setStatus(javax.servlet.http
                        .HttpServletResponse.SC_PARTIAL_CONTENT);
                rangeBytes = request.getHeader("Range").replaceAll("bytes=", "");
                if (rangeBytes.indexOf('-') == rangeBytes.length() - 1) {
                        rangeSwitch = 1;
                        rangeBytes = rangeBytes.substring(0, rangeBytes.indexOf('-'));
                        compeleteLength = Long.parseLong(rangeBytes.trim());
                        contentLength = totalLength - compeleteLength;
                } else {
                        rangeSwitch = 2;
                        String temp0 =
                                rangeBytes.substring(0, rangeBytes.indexOf('-'));
                        String temp2 = rangeBytes.substring(
                                rangeBytes.indexOf('-') + 1,
                                        rangeBytes.length());
                        compeleteLength = Long.parseLong(temp0.trim());
                        toLength = Long.parseLong(temp2);
                        contentLength = toLength - compeleteLength;
                }
        } else {
                contentLength = totalLength;
        }

        response.reset();
```

```
response.setHeader("Accept-Ranges", "bytes");

if (compeleteLength != 0) {
        System.out.println("开始断点续传");
        switch (rangeSwitch) {
        case 1:
                String contentRange1 = new StringBuffer("bytes ")
                        .append(new Long(compeleteLength).toString())
                        .append("-").append(new Long(totalLength - 1).toString())
                        .append("/").append(new Long(totalLength).toString())
                                .toString();
                response.setHeader("Content-Range", contentRange1);
                break;

        case 2:
                String contentRange2 = rangeBytes + "/"
                                + new Long(totalLength).toString();
                response.setHeader("Content-Range", contentRange2);
                break;
        }
} else {
        System.out.println("开始普通下载");
}

try {
        response.addHeader("Content-Disposition", "attachment;
                filename=\""+ downloadFile.getName() + "\"");
        response.addHeader("Content-Length",
                String.valueOf(contentLength));
        response.setContentType("application/octet-stream");
        raf = new RandomAccessFile(downloadFile, "r");
        os = response.getOutputStream();
        bufferOs = new BufferedOutputStream(os);
        try {
                switch (rangeSwitch) {
                case 0:
                case 1:
                        raf.seek(compeleteLength);
                        int n1 = 0;
                        while ((n1 = raf.read(b, 0, 1024)) != -1) {
```

```
                                bufferOs.write(b, 0, n1);
                            }
                            break;

                    case 2:
                            raf.seek(compeleteLength);
                            int n2 = 0;
                            long readLength = 0;
                            while (readLength <= contentLength - 1024) {
                                    n2 = raf.read(b, 0, 1024);
                                    readLength += 1024;
                                    bufferOs.write(b, 0, n2);
                            }
                            if (readLength <= contentLength) {
                                    n2 = raf.read(b, 0, (int) (contentLength - readLength));
                                    bufferOs.write(b, 0, n2);
                            }
                            break;
                    }
                    bufferOs.flush();
                    System.out.println("下载结束");
            } catch (IOException ie) {
            }
        } catch (Exception e) {
            e.printStackTrace();
        } finally {
            try {
                if (bufferOs != null) {
                    bufferOs.close();
                }
                if (raf != null) {
                    raf.close();
                }
            } catch (IOException e) {
                e.printStackTrace();
            }
        }
    }
}
```

上述代码主要实现了断点续传功能中服务器端的代码，具体实现方法在此不再过多讲解。

2．客户端的实现

(1) 客户端的实现相对比较复杂，首先需要创建用于描述下载文件信息的实体类DownloadInfo.java，该类主要用于描述下载文件的名称、已完成的长度和文件总长度，代码如下：

```
public class DownloadInfo {
        private int id;
        private String fileName;
        // 已完成长度
        private int compeleteLength;
        // 文件总长度
        private int totalLength;
        public DownloadInfo() {
        }

        public DownloadInfo(String fileName, int compeleteLength,
        int totalLength) {
                super();
                this.fileName = fileName;
                this.compeleteLength = compeleteLength;
                this.totalLength = totalLength;
        }

        // get()与 set()方法略

}
```

(2) 下载过程中需要将当前下载的进度信息实时地更新到数据库进行保存，以便停止下载后，继续下载剩余部分，因此需要创建数据库操作类“DBHelper.java”，代码如下：

```
public class DBHelper extends SQLiteOpenHelper {

        public DBHelper(Context context) {
                super(context, "download.db", null, 1);
        }

        @Override
        public void onCreate(SQLiteDatabase db) {
                db.execSQL("create table download_info("
                          + "_id integer PRIMARY KEY AUTOINCREMENT, "
                          + "file_name text,"
                          + "compelete_length integer,"
```

```
                + "total_length integer "
                + ")");
    }

    @Override
    public void onUpgrade(SQLiteDatabase db, int oldVersion,
    int newVersion) {          }

    /**
     * 根据文件名称获取文件信息
     *
     * @param fileName
     * @return
     */
    public DownloadInfo getInfoByName(String fileName) {
            SQLiteDatabase db = getReadableDatabase();
            String sql = "select * from download_info where file_name=?";
            Cursor cursor = db.rawQuery(sql, new String[] { fileName });
            DownloadInfo info = null;
            if (cursor.moveToFirst()) {
                    info = new DownloadInfo();
                    info.setId(cursor.getInt(cursor.getColumnIndex("_id")));
                    info.setFileName(cursor.getString(cursor.getColumnIndex("file_name")));
                    info.setCompeleteLength(cursor.getInt(cursor.getColumnIndex("compelete_length")));
                    info.setTotalLength(cursor.getInt(cursor.getColumnIndex("total_length")));
            }
            cursor.close();
            db.close();
            return info;
    }

    /**
     *添加新的下载信息
     *
     * @param info
     */
    public void addNewInfo(DownloadInfo info) {
            SQLiteDatabase db = getWritableDatabase();
            String sql = "insert into download_info(
                    file_name,compelete_length,total_length) values (?,?,?)";
```

```
            Object[] args = { info.getFileName(), info.getCompeleteLength(),info.getTotalLength() };
            db.execSQL(sql, args);
            db.close();
    }

    /**
     * 更新文件进度
     *
     * @param fileName
     *              文件名
     * @param compeleteLength
     *              完成进度
     */
    public void updateProgress(String fileName, int compeleteLength) {
            SQLiteDatabase db = getReadableDatabase();
            String sql = "update download_info set compelete_length=?"
                            + " where file_name=?";
            Object[] args = { compeleteLength, fileName };
            db.execSQL(sql, args);
            db.close();
    }

    /**
     * 删除指定文件信息
     *
     * @param fileName
     */
    public void delete(String fileName) {
            SQLiteDatabase db = getReadableDatabase();
            db.delete("download_info", "file_name=?",new String[] { fileName });
            db.close();
    }
}
```

上述代码中，分别创建了 getInfoByName()、addNewInfo()、updateProgress()、delete()方法来维护下载信息。

(3) 需要创建下载文件的核心类“FileDownloader.java”，该类用于完成整个断点续传下载文件的操作，主要流程如下：

① 使用要下载的文件 URL、文件名称、用于更新下载进度的 Handler 对象创建 FileDownloader 对象。

② 通过刚创建 FileDownloader 对象获取文件下载信息，期间会检查文件是否存在，

如果不存在，则将创建新的文件信息。

③ 通过文件信息初始化用于更新下载进度的 Handler 对象，准备更新进度条。

④ 之后执行下载命令，进行文件的下载。

⑤ 在下载文件的过程中，将每次下载的进度发送到 Handler 对象，同时，更新数据库中对应文件信息的进度。

⑥ 文件完全下载完毕后，将下载信息从数据库中删除。

接下来将分步骤实现 FileDownloader.java 类的功能，首先编写基础代码如下：

```
public class FileDownloader {
    private Context context = null;
    /** 下载状态 - 初始化 */
    public static final int TAG_STATE_INIT = 1;
    /** 下载状态 - 下载中 */
    public static final int TAG_STATE_DOWNLOADING = 2;
    /** 下载状态 - 暂停 */
    public static final int TAG_STATE_STOP = 3;
    /** 更新进度 */
    public static final int TAG_UPDATE_PROGRESS = 101;
    /** 下载状态 */
    private int downloadState = TAG_STATE_INIT;
    /** 下载地址 */
    private String downloadUrl;
    /** 文件名 */
    private String fileName;
    /** 本地文件保存路径 */
    private static String localFilePath;
    /** 文件信息 */
    private DownloadInfo fileInfo = null;
    /** 用于更新进度的 Handler */
    private Handler handler;

    static {
        localFilePath = Environment.getExternalStorageDirectory()
                        .getAbsolutePath() + File.separatorChar;
    }
    /**
     * 构造方法，初始化 FileDownloader
     *
     * @param context
     * @param downloadUrl 下载文件的 URL
     * @param filename 文件名称
```

```
     * @param handler 处理下载进度的 Handler
     */
    public FileDownloader(Context context, String downloadUrl,
            String fileName,Handler handler) {

            this.context = context;
            this.fileName = fileName;
            this.downloadUrl = downloadUrl;
            this.handler = handler;
    }
}
```

上述代码主要定义了下载文件用到的基本属性、用于初始化 FileDownloader 对象的构造方法和 static 静态块中获取手机存储卡的绝对目录的路径。

继续添加 getDownloaderInfo()方法，该方法用于获取下载信息。该方法首先查询数据库中是否存在目标文件的下载记录，如果存在，则将直接使用该条记录；如果不存在，则将调用 createNewFileInfo()方法创建新的文件下载信息，包括在存储卡中创建临时文件、在数据库中添加下载信息。getDownloaderInfo()方法和 createNewFileInfo()方法代码如下：

```
/**
 * 获取下载信息
 */
public DownloadInfo getDownloaderInfo() {
    // 获取保存在数据库中的信息
    DBHelper helper = new DBHelper(context);
    fileInfo = helper.getInfoByName(fileName);

    // 如果数据库中没有记录，则创建新的下载信息
    if (fileInfo == null) {
        fileInfo = createNewFileInfo();
    }
    return fileInfo;
}
/**
 * 创建新的下载信息(本地创建文件，数据库中添加记录)
 */
private DownloadInfo createNewFileInfo() {
    DownloadInfo info = null;
    // 获取服务器上将要下载的文件长度
    int fileLength = getFileLength();
    try {
        if (fileLength > 0) {
```

```
                // 在本地创建文件
                File file = new File(localFilePath + fileName);
                if (file.exists()) {
                        file.delete();
                }
                file.createNewFile();
                RandomAccessFile accessFile = new RandomAccessFile(file, "rwd");
                accessFile.setLength(fileLength);
                accessFile.close();
            }
        } catch (IOException e) {
                e.printStackTrace();
        }
        // 创建下载信息对象
        info = new DownloadInfo(fileName, 0, fileLength);
        // 在数据库中添加记录
        DBHelper helper = new DBHelper(context);
        helper.addNewInfo(info);
        return info;
}
```

创建新的下载信息时，需要获取目标文件的长度，文件长度需要访问服务器获取。getFileLength()方法代码如下：

```
/**
 * 获取服务器文件长度
 */
private int getFileLength() {
        int fileSize = -1;
        try {
                URL url = new URL(downloadUrl);
                HttpURLConnection connection = (HttpURLConnection) url.openConnection();
                connection.setConnectTimeout(5000);
                connection.setRequestMethod("GET");
                fileSize = connection.getContentLength();
                connection.disconnect();
        } catch (IOException e) {
                e.printStackTrace();
        }
        return fileSize;
}
```

该方法中，通过 GET 方式请求服务器，URL 连接是要下载的文件的链接，通过

getContentLength()方法获取文件长度。

至此，下载文件前的准备工作相关方法已编写完成，接下来编写下载文件相关方法，本类中，通过对外界公开的 download()方法执行下载文件任务，实际下载过程是在新的线程中完成的。在下载文件过程中，不断将最新的下载进度等信息更新到数据库，同时将这些信息通过 Handler 发送到相关 Activity 进行处理。该部分代码如下：

```
/**
 * 开始下载文件
 */
public void download() {
        if (downloadState == TAG_STATE_DOWNLOADING) {
                return;
        }
        downloadState = TAG_STATE_DOWNLOADING;
        new DownLoadThread().start();
}
/**
 * 下载文件的线程
 */
private class DownLoadThread extends Thread {
        private int compeleteLength = fileInfo.getCompeleteLength();
        private int totalLength = fileInfo.getTotalLength();

        @Override
        public void run() {
                HttpURLConnection connection = null;
                RandomAccessFile randomAccessFile = null;
                InputStream is = null;

                DBHelper helper = new DBHelper(context);
                try {
                        URL url = new URL(downloadUrl);
                        connection = (HttpURLConnection) url.openConnection();
                        connection.setConnectTimeout(5000);
                        connection.setRequestMethod("GET");
                        // 设置下载范围
                        connection.setRequestProperty("Range", "bytes="
                                        + compeleteLength + "-" + totalLength);
                        randomAccessFile = new RandomAccessFile(localFilePath
                                        + fileName, "rwd");
                        randomAccessFile.seek(compeleteLength);
```

```
                // 将要下载的文件写到保存在保存路径下的文件中
                is = connection.getInputStream();
                // 设置缓冲大小
                byte[] buffer = new byte[1024 * 1024];
                int length = -1;
                while ((length = is.read(buffer)) != -1) {
                    // 停止下载
                    if (downloadState == TAG_STATE_STOP) {
                        return;
                    }
                    randomAccessFile.write(buffer, 0, length);
                    compeleteLength += length;
                    // 更新数据库中的下载信息
                    helper.updateProgress(fileName, compeleteLength);
                    // 用消息将下载信息传给进度条，对进度条进行更新
                    Message msg = Message.obtain();
                    msg.what = TAG_UPDATE_PROGRESS;
                    msg.arg1 = length;
                    handler.sendMessage(msg);
                    // 下载完成
                    if (compeleteLength >= fileInfo.getTotalLength()) {
                        downloadState = TAG_STATE_STOP;
                        // 下载完成后，删除数据库中的记录
                        helper.delete(fileName);
                        return;
                    }
                }
            } catch (Exception e) {
                e.printStackTrace();
            } finally {
                try {
                    is.close();
                    randomAccessFile.close();
                    connection.disconnect();
                } catch (Exception e) {
                    e.printStackTrace();
                }
            }
        }
}
```

该类还向外界提供了两个公有方法，用于获取下载状态和停止下载的操作，代码如下：

```
/**
 * 获取下载状态
 */
public int getDownloadState() {
        return downloadState;
}
/**
 * 停止下载
 */
public void stopDownload() {
        downloadState = TAG_STATE_STOP;
}
```

至此，FileDownloader 工具类已实现完毕。

(4) 实现 MainActivity.java 的编码。首先修改布局文件“activity_main.xml”，代码如下：

```
<RelativeLayout xmlns:android="http://schemas.android.com/apk/res/android"
    xmlns:tools="http://schemas.android.com/tools"
    android:layout_width="match_parent"
    android:layout_height="match_parent" >

<ProgressBar
    android:id="@+id/act_main_progressbar"
    style="?android:attr/progressBarStyleHorizontal"
    android:layout_width="fill_parent"
    android:layout_height="wrap_content"
    android:layout_marginTop="30dp" />

<LinearLayout
    android:layout_width="wrap_content"
    android:layout_height="wrap_content"
    android:layout_below="@id/act_main_progressbar"
    android:layout_centerHorizontal="true"
    android:layout_marginTop="30dp"
    android:orientation="horizontal" >
<Button
    android:id="@+id/act_main_start_btn"
    android:layout_width="wrap_content"
    android:layout_height="wrap_content"
```

```
    android:text="下载" />
<Button
    android:id="@+id/act_main_stop_btn"
    android:layout_width="wrap_content"
    android:layout_height="wrap_content"
    android:layout_marginLeft="30dp"
    android:text="停止" />
</LinearLayout>
</RelativeLayout>
```

上述代码中，在页面中添加了一个 ProgressBar 进度条，用于显示下载文件的进度；添加两个按钮，分别是“下载”和“停止”。

(5) 实现“MainActivity.java”的编码，代码如下：

```
public class MainActivity extends Activity {
    /** 初始化进度条 */
    protected static final int TAG_INIT_PROGRESS = 201;
    private ProgressBar progressBar = null;
    private Button startBtn = null;
    private Button stopBtn = null;

    FileDownloader downloader = null;
    DownloadInfo info;

    @Override
    protected void onCreate(Bundle savedInstanceState) {
        super.onCreate(savedInstanceState);
        setContentView(R.layout.activity_main);

        String fileName = "music.mp3";
        String downloadUrl = "http://192.168.1.48:8080/HttpService/DownloadServlet";
        // 创建下载器
        downloader = new FileDownloader(this, downloadUrl, fileName, handler);

        // 初始化控件
        progressBar = (ProgressBar) findViewById(R.id.act_main_progressbar);
        startBtn = (Button) findViewById(R.id.act_main_start_btn);
        stopBtn = (Button) findViewById(R.id.act_main_stop_btn);

        // 添加事件监听
        startBtn.setOnClickListener(new OnClickListener() {
            @Override
```

```
                public void onClick(View v) {
                        startDownload();
                }
        });

        stopBtn.setOnClickListener(new OnClickListener() {
                @Override
                public void onClick(View v) {
                        Toast.makeText(getApplicationContext(), "停止下载",
                                        Toast.LENGTH_SHORT).show();
                        downloader.stopDownload();
                }
        });
}

/**
 * 开始下载
 */
protected void startDownload() {
        new Thread(new Runnable() {

                @Override
                public void run() {
                        // 获取下载信息
                        info = downloader.getDownloaderInfo();
                        // 通知 handler 初始化进度条
                        handler.obtainMessage(TAG_INIT_PROGRESS,
                                info.getCompeleteLength(),
                                info.getTotalLength()).sendToTarget();
                        // 开始下载文件
                        downloader.download();
                }
        }).start();

}

/**
 * 更新下载的进度
 */
private Handler handler = new Handler() {
```

```
            public void handleMessage(Message msg) {

                switch (msg.what) {
                case TAG_INIT_PROGRESS:
                    progressBar.setMax(msg.arg2);
                    progressBar.setProgress(msg.arg1);
                    break;

                case FileDownloader.TAG_UPDATE_PROGRESS:
                    int length = msg.arg1;// 每次下载的大小
                    progressBar.incrementProgressBy(length);
                    if (progressBar.getProgress() >=
                        info.getTotalLength()) {
                        Toast.makeText(getApplicationContext(),
                            "下载完毕",Toast.LENGTH_SHORT).show();
                    }
                    break;
                }
            }
        };
}
```

上述代码中，点击“下载”按钮，开始下载文件，如果之前已经下载了一部分，会首先将进度条初始化到指定的位置，下载过程中会不断更新进度条，下载完毕后，会以 Toast 的方式提示“下载完毕”。下载过程中单击“停止”按钮，将停止下载，同时会以 Toast 的方式提示“停止下载”。

(6) 在“AndroidManifest.xml”配置文件中添加权限，代码如下：

```
<!--访问网络权限 -->
<uses-permission android:name="android.permission.INTERNET"/>
<!-- SD 卡写入权限 -->
<uses-permission android:name=
    "android.permission.WRITE_EXTERNAL_STORAGE"/>
```

(7) 检查运行结果，启动服务器和客户端，点击“下载”按钮，下载完成后打开存储卡，查看下载的音乐是否能正常播放。

本 章 小 结

(1) HttpURLConnection 继承自 URLConnection 抽象类，也是一个抽象类。

(2) HttpURLConnection 对象需要通过调用 URL 对象的 openConnection()方法获取。

(3) HttpClient 是由 Apache 提供的 HTTP 客户端组件，比较适合在 Android 中进行网络应用开发。

(4) HttpClient 中，通过 HttpPost 向服务器发送 Post 请求，通过 HttpGet 发送 Get 请求。
(5) 上传文件到服务器以及断点续传下载文件需要有服务器的支持才能实现。
(6) 断点续传下载文件，相关断点数据通常使用数据库进行管理。

本 章 练 习

(1) 在 Android 中实现网络请求，可以使用__________或__________两个类来实现。
(2) 编写程序，实现头像上传功能(将本地图片上传到服务器)。
(3) 编写程序，实现将服务器中的文件通过断点续传技术下载到本地。

第4章　高级用户体验

本章目标

- 理解图片自适应原理
- 掌握 Draw9-patch 工具的使用
- 掌握 ListView 基本适配器的使用
- 理解 ListView 自定义适配器的使用
- 理解 PopupWindow 的简单使用

4.1 图片自适应

目前市面上的 Android 手机各式各样，因此屏幕分辨率也各不相同，这就意味着同一套 UI 切图需要尽量匹配具有多种不同分辨率的屏幕，某些极端情况下，会为每一个常见的屏幕分辨率制作一套对应的 UI 切图，这对于 Android 开发者和美工来说也是一件比较头痛的事情。针对以上情况，Android SDK 中为开发者提供了一个处理图片的工具来解决这一问题，这个工具名叫“Draw9-patch”。本节将重点讲解此工具的使用。

4.1.1 Draw9-patch 概述

“Draw9-patch”工具的功能是可以指定图片中哪些区域可以进行拉伸或缩放，哪些区域是固定不变的。这样一来，可以有效避免某些图片因手机分辨率不同发生失真或变形的问题。生成的图片文件格式为“.9.png”，它是一种用于 Android 程序的特殊的图片格式，当 Android 遇到这种格式的图片时，会进行特殊的处理。当然，实际上还是“.png”格式的图片。

“Draw9-patch”工具位于 Android SDK 目录下的“tools”文件夹中，文件名称为“draw9patch.bat”。它是一个批处理文件，双击图标后将自动启动“Draw9-patch”工具，运行后的界面如图 4-1 所示。

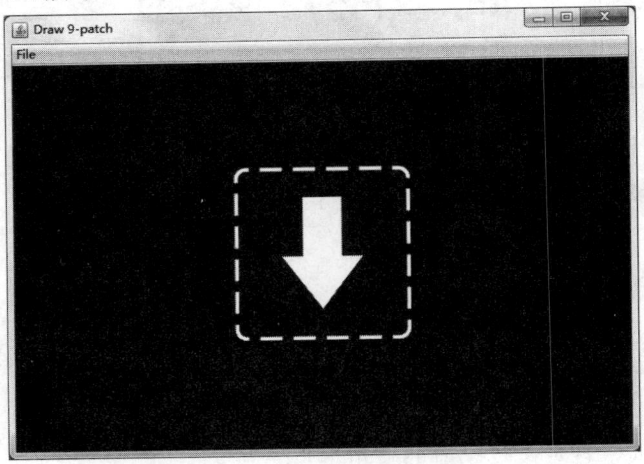

图 4-1　Draw9-patch 工具

可以将要进行处理的图片直接拖动到界面中打开，也可以通过“File”->“Open 9-patch...”菜单命令打开图片。打开图片后的软件界面如图 4-2 所示。

界面主要分为三大部分：

- ✧ 左侧部分为编辑区，可以控制图片在水平或垂直缩放时，固定不变的部分和自动缩放部分以及内容填充部分。
- ✧ 右侧部分为预览区，此区域有三种预览效果，从上到下依次为：纵向拉伸、横向拉伸、双向拉伸；

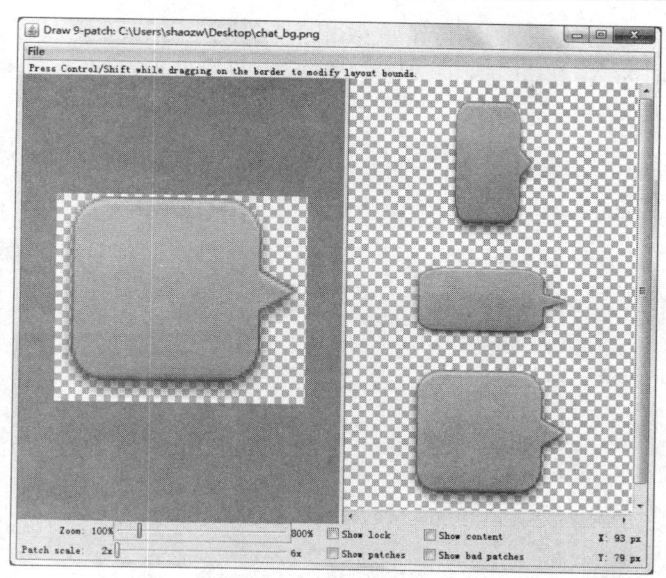

图 4-2　Draw9-patch 工具处理图片前

✧ 底部部分为选项区，主要是对以上两个区域的辅助控制，其中：
  ➢ Zoom：编辑区的缩放。
  ➢ Patch scale：作用于预览区，控制预览缩放。
  ➢ Show lock：显示不可绘区域，用红色粗斜线区域表示(把鼠标移动到图片上方即可显示)。
  ➢ Show patches：显示可延伸区域，用紫色区域表示。
  ➢ Show content：作用于预览区，显示内容填充区域。
  ➢ Show bad patches：如果设计的伸缩区域不合理，会以显示红框的形式作出提示，以尽可能防止错误的发生。

打开一张图片后，工具会自动在图片四周各增加一个像素的空间，这块空间用于绘制图片的拉伸区域、内容填充区域的规则线(点)，图片本身属于不可编辑区域。绘制这些规则线(点)的方式很简单，在需要的位置点击或拖动即可，按住“Shift”键再次点击或拖动便可取消绘制。

“.9.png”图片的规则如下：

✧ 图片四周最外侧边缘 1 像素的区域为规则线区域；
✧ 左边缘和上边缘分别用于规定图片水平方向和垂直方向拉伸规则，这两条规则必须存在；
✧ 右边缘和下边缘用于规定图片内容水平方向和垂直方向填充区域规则，可理解为同“padding”属性的作用，这两条规则可选。

4.1.2　绘制图片缩放

下述示例用于实现：使用“Draw9-patch”工具，对“chat_bg_old.png”图片进行修改，使图片能够自动缩放而不影响显示效果，将修改后的图片保存为“chat_bg_new.9.png”，然后创建新的 Android 项目，对修改前后的图片进行对比。

1．制作自适应图片

在“Draw9-patch”工具中，打开图片“chat_bg_old.png”，绘制水平和垂直缩放规则，并勾选“Show patches”，显示可延伸区域，如图 4-3 所示。

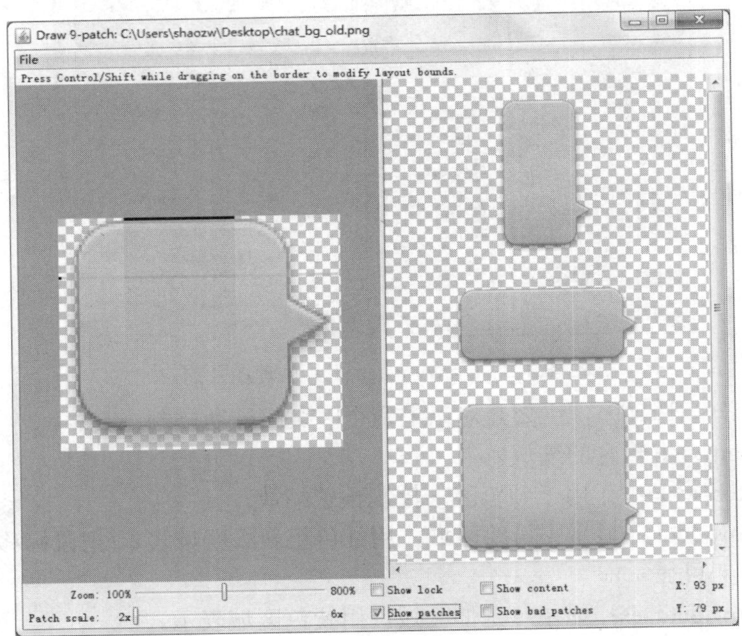

图 4-3　Draw9-patch 工具处理图片后

绘制规则完成后，可以查看右侧的预览区，确认无误后，通过单击“Ctrl+S”键或者通过菜单命令将图片保存为“chat_bg_new.9.png”。

2．进行效果对比

新建 Android 项目，将“chat_bg_old.png”和“chat_bg_new.9.png”一并拷贝到项目的“drawable-hdpi”资源文件夹中。

编写“activity_main.xml”布局文件，添加两个 TextView 文本控件，分别设置不同的背景图片，代码如下：

```
<LinearLayout xmlns:android="http://schemas.android.com/apk/res/android"
    xmlns:tools="http://schemas.android.com/tools"
    android:layout_width="match_parent"
    android:layout_height="match_parent"
    android:background="#ededed"
    android:orientation="vertical"
    android:padding="20dp" >

<TextView
```

```
    android:layout_width="wrap_content"
    android:layout_height="wrap_content"
    android:background="@drawable/chat_bg_old"
    android:text="床前明月光，\n 疑是地上霜；\n 举头望明月，\n 低头思故乡。\n(未处理之前的原始图
片)" />

<TextView
    android:layout_width="wrap_content"
    android:layout_height="wrap_content"
    android:layout_marginTop="30dp"
    android:background="@drawable/chat_bg_new"
    android:text="床前明月光，\n 疑是地上霜；\n 举头望明月，\n 低头思故乡。\n(进行缩放规则处理后的
图片)" />

</LinearLayout>
```

上述代码中，在页面中添加了两个 TextView 文本控件，位于上方的 TextView 设置的背景图片为修改之前的“chat_bg_old.png”，位于下方的 TextView 设置的背景图片为修改之后的“chat_bg_new.9.png”。运行程序，对比效果如图 4-4 所示。很明显，上方的 TextView 背景图已经失真变形，而下方的 TextView 背景图并没有失真变形。

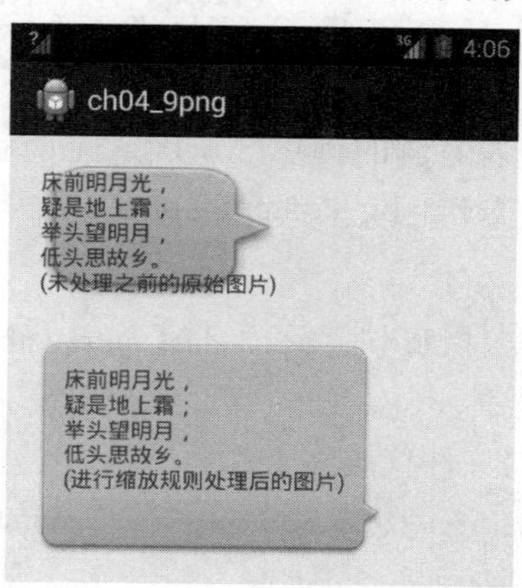

图 4-4　图片处理前后的效果对比

4.1.3　绘制内容填充区域

在 4.1.2 节的示例中，读者会发现，虽然实现了图片的正常缩放，但是总感觉图片里面的内容有些不和谐。下述示例用于实现：在“chat_bg_new.9.png”图片的基础上，绘制内容填充区域规则，将生成的文件命名为“chat_bg_content.9.png”，与 4.1.2 小节的示例进行对比。

1．为图片绘制填充区域规则

绘制之前需要注意的是，如果 4.1.2 小节中打开的“Draw9-patch”工具没有关闭，还

处于制作“chat_bg_new.9.png”图片的状态，则可以直接进行绘制；如果已经关闭，则需要重新打开原始的“chat_bg_old.png”图片，重新绘制缩放规则和内容填充区域规则。勾选“Show content”选项，预览区显示填充区域，结果如图 4-5 所示。

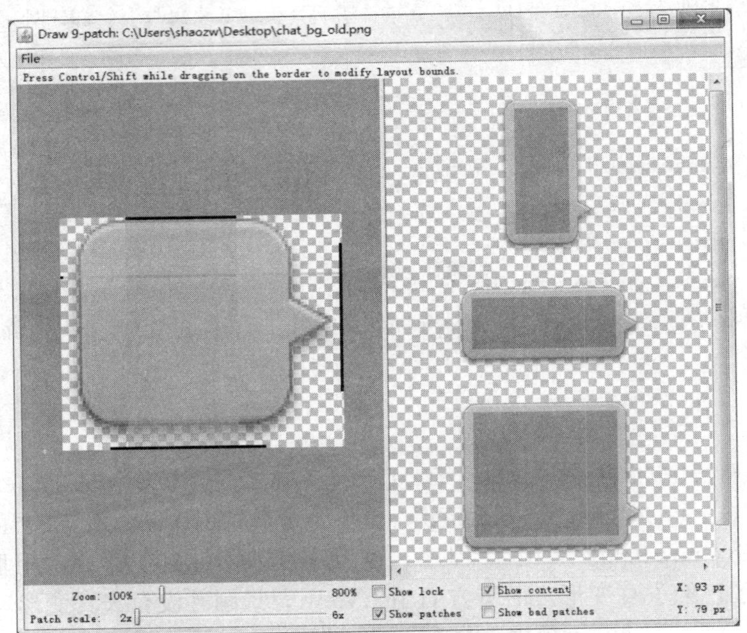

图 4-5　绘制内容填充区域

绘制规则完成后，可以查看右侧的预览区，确认无误后，通过单击 “Ctrl+S”键或者通过菜单命令将图片保存为“chat_bg_content.9.png”。

2. 进行效果对比

将“chat_bg_content.9.png”拷贝到之前项目的“drawable-hdpi”资源文件夹中，修改“activity_main.xml”布局文件，添加一个 TextView 文本控件，将背景图设为“chat_bg_content.9.png”，布局文件代码如下：

```
<LinearLayout xmlns:android="http://schemas.android.com/apk/res/android"
    xmlns:tools="http://schemas.android.com/tools"
    android:layout_width="match_parent"
    android:layout_height="match_parent"
    android:background="#ededed"
    android:orientation="vertical"
    android:padding="20dp" >

<TextView
    android:layout_width="wrap_content"
    android:layout_height="wrap_content"
    android:background="@drawable/chat_bg_old"
```

```
    android:text="床前明月光，\n 疑是地上霜；\n 举头望明月，\n 低头思故乡。\n(未处理之前的原始图
片)" />

<TextView
    android:layout_width="wrap_content"
    android:layout_height="wrap_content"
    android:layout_marginTop="30dp"
    android:background="@drawable/chat_bg_new"
    android:text="床前明月光，\n 疑是地上霜；\n 举头望明月，\n 低头思故乡。\n(进行缩放规则处理后的
图片)" />

<TextView
    android:layout_width="wrap_content"
    android:layout_height="wrap_content"
    android:layout_marginTop="30dp"
    android:background="@drawable/chat_bg_content"
    android:text="床前明月光，\n 疑是地上霜；\n 举头望明月，\n 低头思故乡。\n(增加了内容填充区域规
则的图片)" />

</LinearLayout>
```

上述代码中，在之前的基础上增加了一个 TextView 文本控件，背景图为“chat_bg_content.9.png”，运行项目后，与之前的两个 TextView 文本控件进行效果对比，会发现增加了内容填充区域规则的“.9.png”图片达到了预期效果，效果如图 4-6 所示。

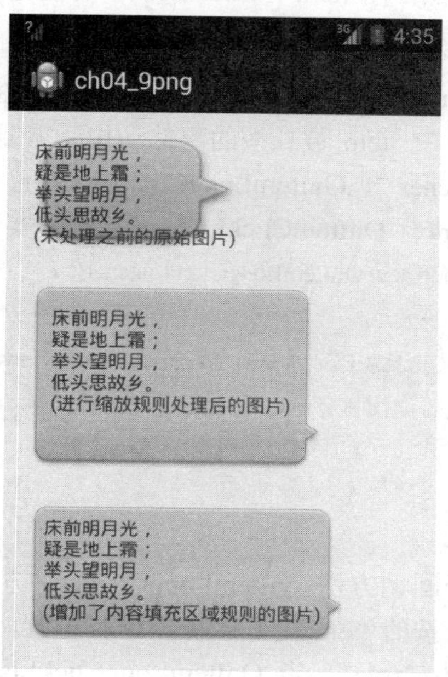

图 4-6　修改图片填充区域前后效果对比

4.2 ListView 列表视图

ListView 称为列表视图，主要用于显示一系列相似的数据，当数据过多，超出屏幕时，可以通过纵向滑动的方式查看其他部分，ListView 的使用非常普遍，几乎在每一个 Android 程序中都有 ListView 的身影。

4.2.1 ListView 事件处理

ListView 控件支持与用户交互，包括“滑动”“点击”“长按”等操作，常用的事件监听器有以下几种。

1．OnItemClickListener

OnItemClickListener 监听器实现比较简单，主要用于监听 ListView 中 Item 被点击的事件动作，具体实现代码如下：

```
listView.setOnItemClickListener(new OnItemClickListener() {
    @Override
    public void onItemClick(AdapterView<?> parent, View view,int position, long id) {
    }
});
```

该监听器中实现了一个回调方法 onItemClick()，方法中的参数通常只需要关心“position”参数，其表示被点击的 Item 在 ListView 中的位置，从 0 开始。

2．OnItemLongClickListener

OnItemLongClickListener 监听器与 OnItemClickListener 监听器类似，唯一的区别就是该监听器用于监听 ListView 中 Item 被长按的事件动作。需要注意的是，如果对 ListView 同时设置了 OnItemClickListener 和 OnItemLongClickListener，当 OnItemLongClickListener 被触发执行后，通常会接着执行 OnItemClickListener。具体实现代码如下：

```
listView.setOnItemLongClickListener(new OnItemLongClickListener() {
    @Override
    public boolean onItemLongClick(AdapterView<?> parent, View view,int position, long id) {
        // 返回值将决定是否继续执行 OnItemClickListener
        return false;
    }
});
```

该监听器中实现了一个回调方法 onItemLongClick()，方法中的参数通常只需要关心“int position”，其表示被长按的 Item 在 ListView 中的位置，从 0 开始。该方法返回值为 boolean 类型，如果返回值为“true”，当 OnItemLongClickListener 执行完毕后，将不再执行 OnItemClickListener 监听器；否则，将继续执行 OnItemClickListener 监听器。

3．OnScrollListener

虽然 OnScrollListener 相比前两个监听器较为复杂，但比较实用，主要用于数据分页显示的实现、监听 ListView 滑动状态等，具体实现代码如下：

```
listView.setOnScrollListener(new OnScrollListener() {
    @Override
    public void onScrollStateChanged(AbsListView view, int scrollState) {
    }

    @Override
    public void onScroll(AbsListView view, int firstVisibleItem,
                int visibleItemCount, int totalItemCount) {
    }
});
```

该监听器实现了两个回调方法。

✧ onScrollStateChanged()：当 ListView 滑动状态改变时触发回调，参数“int scrollState”通常有以下三种状态。
  ➢ SCROLL_STATE_IDLE：ListView 停止滚动，值为 0。
  ➢ SCROLL_STATE_TOUCH_SCROLL：当 ListView 正在滑动并且用户的手指还在屏幕上时，值为 1。
  ➢ SCROLL_STATE_FLING：当用户的手指离开屏幕之前快速滑动了 ListView，由于惯性 ListView 继续滑动时，值为 2。

✧ onScroll()：当 ListView 滑动时触发回调，主要参数如下。
  ➢ firstVisibleItem：int 类型，表示 ListView 中当前可见的 Item 的索引位置，值从 0 开始。
  ➢ visibleItemCount：int 类型，表示 ListView 中当前可见的 Item 的数量。
  ➢ totalItemCount：int 类型，表示当前与 ListView 绑定的数据的总条数。

4.2.2 Adapter 概述

要想实现 ListView 的数据显示，仅使用 ListView 控件是不够的，还需要其他两个必要的元素，那就是 Adapter(适配器)和数据源。

1．Adapter 分类

通常情况下，ListView 中每个条目(Item)都是复用同一个布局，但有时为了显示更为复杂的效果，可能会复用多个布局。那么，如何组织这些条目的显示方式，如何将数据绑定到 ListView 并显示，这就需要 Adapter(适配器)类。作为一个中间件，适配器负责数据与 ListView 的绑定。

适配器相关的类主要有四个，这四个类均是 Adapter 类的子类：

✧ ArrayAdapter：最简单的一个 Adapter，为 ListView 提供简单的布局，用于单纯的显示一段文本。

- ✧ SimpleCursorAdapter：通常用于直接显示数据库查询结果等，比较单一，通常不会使用。
- ✧ SimpleAdapter：相对于 ArrayAdapter 较为复杂，但具有良好的扩展性，因为它允许开发者自行创建布局文件进行绑定，可以自定义显示效果。
- ✧ BaseAdapter：是一个抽象类，也是以上三个 Adapter 的基类，在实际开发中，通常会通过继承此类创建自定义的 Adapter 来实现更为复杂的功能，同样需要开发者自行创建布局文件进行绑定，也是使用最频繁的一个 Adapter。

2．Adapter 基本实现原理

通常，绑定到 ListView 的 Adapter 类都需要重写下列方法。

- ✧ getCount()：获取数据源的总条数。
- ✧ getItem(int position)：获取指定位置的对象。
- ✧ getItemId(int position)：获取指定位置 Item 的 id。
- ✧ getView(int position, View convertView, ViewGroup parent)：生成要显示的 View 并提交显示，参数解释如下：
  - ➢ int position：当前所绘制的 Item 的位置，从 0 开始。
  - ➢ View convertView：当前所绘制的 Item 的视图。
  - ➢ ViewGroup parent：表示父容器。

Adapter 将数据绑定到 ListView 的执行过程是：首先通过 getCount()方法获取数据源的条数，然后根据数据的条数逐一调用 getView()方法将内容(Item)绘制到 ListView，也就是说，getCount()方法返回值是多少，就会执行多少次 getView()方法，直到所有 Item 绘制完毕，这样就实现了最终的 ListView 列表页面。

自定义的 Adapter 与 Android 提供的 Adapter 不同的是，后者已经对这些方法进行了封装，开发者只需提供数据源和所需的资源文件即可实现简单的 ListView 显示效果，而自定义的 Adapter 需要开发者自行继承 BaseAdapter 并且重写这些方法来实现相应功能。

3．Adapter 数据刷新

当数据源内容发生变化时，需要通过执行 Adapter 的 notifyDataSetChanged()和 notifyDataSetInvalidated()方法进行“刷新”操作。

- ✧ notifyDataSetChanged()：只重新绘制(刷新)当前可见区域，“刷新数据”通常采用此方法。
- ✧ notifyDataSetInvalidated()：重新绘制(刷新)整个 ListView，恢复初始状态。

4.2.3 ArrayAdapter

ArrayAdapter 的使用非常简单，通常用以下构造方法进行创建：

```
ArrayAdapter<T>(Context context, int textViewResourceId, T[] objects)
```

该构造方法中有三个参数：

- ✧ Context context：上下文对象。

✧ int textViewResourceId：引用的布局文件资源 id，通常使用系统提供的。
✧ T[] objects：数据源，任意类型的数组，如果数组中保存的是对象，会调用该对象的“toString()”方法。

该构造方法还有另一种重载的实现方式：

```
ArrayAdapter<T>(Context context, int textViewResourceId, List<T> objects)
```

在数据源方面，该构造方法可以使用 List 集合作为数据源。

下述示例用于实现：使用 ListView 绑定 ArrayAdapter 的方式显示数据。要求：当点击 ListView 中某一条 Item 时，以 Toast 方式提示点击的是第几条。

(1) 新建项目，首先创建 Activity 的布局文件“act_list.xml”，编写代码如下：

```
<RelativeLayout xmlns:android="http://schemas.android.com/apk/res/android"
    xmlns:tools="http://schemas.android.com/tools"
    android:layout_width="match_parent"
    android:layout_height="match_parent"
    android:background="#ededed" >

<ListView
        android:id="@+id/act_list"
        android:layout_width="fill_parent"
        android:layout_height="fill_parent" />

</RelativeLayout>
```

上述代码中，只在布局文件中添加了一个 ListView 控件。

(2) 创建 Activity 类“ArrayAdapterActivity.java”，并绑定布局文件“act_list.xml”，编写代码如下：

```
public class ArrayAdapterActivity extends Activity {

    @Override
    protected void onCreate(Bundle savedInstanceState) {
            super.onCreate(savedInstanceState);
            setContentView(R.layout.act_list);

            // 创建数据源
            String[] strs = new String[20];
            for (int i = 0; i < strs.length; i++) {
                    strs[i] = "第 " + (i + 1) + " 行";
            }

            // 创建 ArrayAdapter
            ArrayAdapter<String> adapter = new ArrayAdapter<String>(this,
                            android.R.layout.simple_list_item_1, strs);
```

```
            ListView listView = (ListView) findViewById(R.id.act_list);
            //为 ListView 绑定适配器
            listView.setAdapter(adapter);

            // 添加点击事件
            listView.setOnItemClickListener(new OnItemClickListener() {

                @Override
                public void onItemClick(AdapterView<?> parent, View view,
                        int position, long id) {
                    Toast.makeText(getApplicationContext(),
                        "你点击了第 " + (position + 1) + " 条",
                        Toast.LENGTH_SHORT).show();
                }
            });
        }
}
```

上述代码中，为 ListView 添加了点击事件，需要注意的是，添加点击事件和绑定 Adapter 没有先后顺序；在点击事件监听器中，因为“position”是从 0 开始的，所以在这里需要“加 1”，表示点击了第几条。

(3) 在 AndroidManifest.xml 文件中对 ArrayAdapterActivity 进行注册，编写代码如下：

```
<activity android:name=".ArrayAdapterActivity" />
```

(4) 运行程序，显示效果如图 4-7 所示。

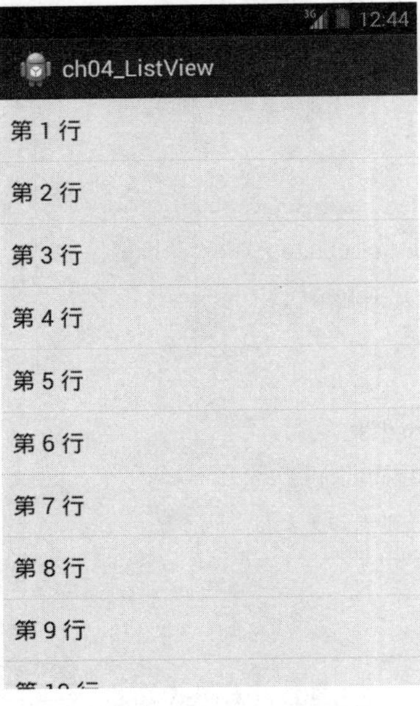

图 4-7　ArrayAdapter 绑定数据

4.2.4 SimpleAdapter

SimpleAdapter 相比 ArrayAdapter 更为复杂，需要用户自行创建 Item 的布局文件，与之绑定的数据源也较为复杂，可使用以下构造方法创建 SimpleAdapter：

```
new SimpleAdapter(Context context, List<? extends Map<String,?>> data, int resource, String[] from, int[] to)
```

该构造方法中有 5 个参数：

✧ Context context：上下文对象。

✧ List<? extends Map<String,?>> data：数据源，这里的数据源比较复杂，是将一系列的 Map 对象保存到 List 列表中，每一个 Map 对象用来表示列表中一个 Item。

✧ int resource：布局文件的资源 id，该布局文件决定了 ListView 中 Item 的显示内容与布局。

✧ String[] from：是一个 String 类型的数组，数组中的值对应数据源中 Map 对象的“KEY”值。

✧ int[] to：所引用的布局文件中控件的 id，顺序必须与“String[] from”参数中的值一一对应。SimpleAdapter 会将数据源中 Map 对象的“VALUE”值自动添加到对应的控件。

下述示例用于实现：使用 ListView 绑定 SimpleAdapter 的方式显示数据。

(1) 将用于显示在列表中的图片复制到项目中的“drawable-hdpi”文件夹。

(2) 创建 ListView 的 Item 布局文件“item_adapter_layout.xml”，代码如下：

```
<?xml version="1.0" encoding="utf-8"?>
<LinearLayout xmlns:android="http://schemas.android.com/apk/res/android"
    android:layout_width="match_parent"
    android:layout_height="match_parent"
    android:background="#ededed"
    android:orientation="horizontal" >

    <ImageView
        android:id="@+id/item_pic_iv"
        android:layout_width="64dp"
        android:layout_height="64dp"
        android:baselineAlignBottom="true"
        android:paddingLeft="8dp" />

    <LinearLayout
        android:layout_width="match_parent"
        android:layout_height="wrap_content"
        android:orientation="vertical" >
```

```
    <TextView
        android:id="@+id/item_name_tv"
        android:layout_width="wrap_content"
        android:layout_height="wrap_content"
        android:paddingLeft="8dp"
        android:textColor="#000000"
        android:textSize="20sp" />

    <TextView
        android:id="@+id/item_introduce_tv"
        android:layout_width="wrap_content"
        android:layout_height="wrap_content"
        android:layout_marginLeft="8dp"
        android:textColor="#4d4d4d"
        android:textSize="14sp" />
</LinearLayout>

</LinearLayout>
```

上述代码中，在布局中添加了一个 ImageView 控件，用于显示头像图片；两个 TextView 控件，分别用于显示名称和描述信息。

(3) 创建 Activity 的布局文件，该文件可以直接复用 4.2.3 小节中的“act_list.xml”，不需要做任何改动。

(4) 创建 Activity 类“SimpleAdapterActivity.java”，并绑定布局文件“act_list.xml”，编写代码如下：

```
public class SimpleAdapterActivity extends Activity {

    private String[] names = new String[] { "鼠","牛","虎","兔","龙","蛇","马","羊","猴","鸡","狗","猪" };
    // 用于显示的图片资源 id
    private int[] picIds = new int[] { R.drawable.icon_rat,
                R.drawable.icon_cow, R.drawable.icon_tiger,
                R.drawable.icon_rabbit,R.drawable.icon_dragon,
                R.drawable.icon_snake,R.drawable.icon_horse,
                R.drawable.icon_sheep,R.drawable.icon_monkey,
                R.drawable.icon_cock, R.drawable.icon_dog,
                R.drawable.icon_pig };

    @Override
    protected void onCreate(Bundle savedInstanceState) {
        super.onCreate(savedInstanceState);
        setContentView(R.layout.act_list);

        // 创建数据源
```

```
            List<Map<String, Object>> data =
                    new ArrayList<Map<String, Object>>();
            for (int i = 0; i < 12; i++) {
                    Map<String, Object> map = new HashMap<String, Object>();
                    map.put("pic", picIds[i]);
                    map.put("name", names[i]);
                    map.put("introduce", "我是十二生肖中的 " + names[i]);
                    data.add(map);
            }

            // 创建 SimpleAdapter
            SimpleAdapter myAdapter =
                    new SimpleAdapter(getApplicationContext(),
                    data, R.layout.item_adapter_layout,
                    new String[] { "pic","name", "introduce" },
                    new int[] { R.id.item_pic_iv, R.id.item_name_tv, R.id.item_introduce_tv });

            ListView listView = (ListView) findViewById(R.id.act_list);
            listView.setAdapter(myAdapter);
    }
}
```

上述代码中，实现了 SimpleAdapter 适配器的使用。在使用时，需要注意数据源的格式，以及创建 SimpleAdapter 对象时，参数的对应关系。

(5) 在 AndroidManifest.xml 文件中对 ArrayAdapterActivity 进行注册，编写代码如下：

```
<activity android:name=".SimpleAdapterActivity" />
```

(6) 运行程序，显示效果如图 4-8 所示。

图 4-8　SimpleAdapter 绑定数据

4.2.5　自定义 Adapter

前文中介绍的几种 Adapter，往往在实际开发中无法达到预期效果，而且数据操作比较困难。这主要表现在：当我们需要为 Item 中的控件添加事件时，这几种 Adapter 是无法实现的。

自定义的 Adapter，一般是通过继承 BaseAdapter 来实现的，能够实现各种复杂的页面布局，编程更加灵活，相对也要复杂一些，尤其是性能优化方面需要多加注意。

下述示例用于实现：使用 ListView 自定义 Adapter 的方式实现 4.2.4 小节中相同的显示效果。要求：当点击 ListView 中某一条 Item 时，以 Toast 方式提示点击的是第几条；当点击头像图片时，以 Toast 方式提示当前所点击的 Item 中的描述信息。

(1) 创建 ListView 的 Item 布局文件，在此重用“item_adapter_layout.xml”即可。

(2) 创建实体类“Animal.java”，用于描述数据源中的对象，编写代码如下：

```
public class Animal {
	private int pic;
	private String name;
	private String introduce;

	public Animal() {
	}

	public int getPic() {
		return pic;
	}

	public void setPic(int pic) {
		this.pic = pic;
	}

	public String getName() {
		return name;
	}

	public void setName(String name) {
		this.name = name;
	}

	public String getIntroduce() {
		return introduce;
	}
```

```
    public void setIntroduce(String introduce) {
        this.introduce = introduce;
    }

}
```

上述代码中，在实体类中创建了三个属性“pic”、“name”、“introduce”，分别表示头像图片对应的资源 id、名称、描述信息。

(3) 创建继承“BaseAdapter”的自定义 Adapter 类“CustomAdapter.java”，并实现相应的方法，编写代码如下：

```
public class CustomAdapter extends BaseAdapter {

    private Context ctx;
    // 数据源
    private List<Animal> datas = null;

    public CustomAdapter(Context ctx, List<Animal> datas) {
        super();
        this.ctx = ctx;
        this.datas = datas;
    }

    @Override
    public int getCount() {
        return datas.size();
    }

    @Override
    public Object getItem(int position) {
        return datas.get(position);
    }

    @Override
    public long getItemId(int position) {
        return 0;
    }

    @Override
    public View getView(int position, View convertView,
     ViewGroup parent) {
```

```
            // 从数据源中取出当前要显示的对象
            final Animal animal = datas.get(position);

            if (convertView == null) {
                // 为了避免重复加载资源影响性能，只有在 contentView 为空时才加载
                convertView = LayoutInflater.from(ctx).inflate(
                        R.layout.item_adapter_layout, parent, false);
            }

            // 引用加载布局文件中的控件
            ImageView picIv = (ImageView) convertView.findViewById(R.id.item_pic_iv);
            TextView nameTv = (TextView) convertView.findViewById(R.id.item_name_tv);
            TextView introTv = (TextView) convertView.findViewById(R.id.item_introduce_tv);

            // 将需要显示的信息添加到控件
            picIv.setImageResource(animal.getPic());
            nameTv.setText(animal.getName());
            introTv.setText(animal.getIntroduce());

            // 为头像图片添加点击事件
            picIv.setOnClickListener(new OnClickListener() {

                @Override
                public void onClick(View arg0) {
                    Toast.makeText(ctx, animal.getIntroduce(),
                            Toast.LENGTH_SHORT).show();
                }
            });

            // 返回(提交)当前处理完毕的 Item 布局
            returnconvertView;
        }
}
```

上述代码中，将数据源和 Context 上下文对象通过构造方法传入，并重写了必要的方法。在 getView()方法中，对头像图片做了特殊处理，添加了点击事件，当点击头像图片时，会以 Toast 形式显示所对应的描述信息。

简单来说，ListView 中显示的每一条数据(Item)都存在两种状态，即可见状态和不可见状态。为了有效地节省系统资源和增加用户体验，Android 会自动对 ListView 进行优化：当 Item 完全不可见时，其会被自动释放并回收；当再次显示时，会重新创建相应的

布局。

当滑动 ListView 时，Item 会不断进行绘制。当前已经显示的 Item 在还没有完全不可见时，也会通过 getView()方法进行绘制，但是它的视图对象是已经被创建过的(此时不为 null)，所以没有必要重新加载一次布局视图。因此，在加载布局视图之前，首先需要对视图对象进行非空判断，只有在其为“null”时才创建，这样能够有效避免重复加载资源而影响性能的情况。

注意：在 Android 提供的控件中，GridView、Spinner 等这些与数据源有关的控件都会用到 Adapter，其使用 Adapter 的方式与 ListView 是相同的，因此不再赘述，感兴趣的读者可以自行尝试。

(4) 创建 Activity 的布局文件，该文件可以直接复用 4.2.3 小节中的“act_list.xm”，不需要做任何改动。

(5) 创建 Activity 类“CustomAdapterActivity.java”，并绑定布局文件“act_list.xml”，编写代码如下：

```
public class CustomAdapterActivity extends Activity {

    private String[] names = new String[] { "鼠", "牛", "虎", "兔", "龙", "蛇", "马","羊", "猴", "鸡", "狗", "
猪" };
    // 用于显示的图片资源 id
    private int[] picIds = new int[] { R.drawable.icon_rat,
            R.drawable.icon_cow, R.drawable.icon_tiger,
            R.drawable.icon_rabbit, R.drawable.icon_dragon,
            R.drawable.icon_snake, R.drawable.icon_horse,
            R.drawable.icon_sheep, R.drawable.icon_monkey,
            R.drawable.icon_cock, R.drawable.icon_dog,
            R.drawable.icon_pig };

    private ListView listView = null;
    // 自定义的 Adapter
    private CustomAdapter adapter = null;

    @Override
    protected void onCreate(Bundle savedInstanceState) {
        super.onCreate(savedInstanceState);
        setContentView(R.layout.act_list);

        // 创建数据源
        List<Animal> data = new ArrayList<Animal>();
        for (int i = 0; i < 12; i++) {
            Animal animal = new Animal();
            animal.setPic(picIds[i]);
```

```
                animal.setName(names[i]);
                animal.setIntroduce("我是十二生肖中的 " + names[i]);
                data.add(animal);
            }

            // 创建 CustomAdapter 对象
            adapter = new CustomAdapter(this, data);

            // 为 ListView 绑定 Adapter
            listView = (ListView) findViewById(R.id.act_list);
            listView.setAdapter(adapter);

            listView.setOnItemClickListener(new OnItemClickListener() {

                @Override
                public void onItemClick(AdapterView<?> parent, View view,
                        int position, long id) {
                    Toast.makeText(getApplicationContext(),
                            "你点击了第 " + (position + 1) + " 条",
                            Toast.LENGTH_SHORT).show();
                }
            });
    }
}
```

上述代码中，数据源类型为 List<Animal>，集合中存放的是一个自定义的实体类。相对于 4.2.3 小节中的数据源 List<Map<String, Object>>类型，从实际商业项目角度来看，更符合开发需要，原因在于：当前程序开发中，客户端和服务器进行交互时，数据的格式通常为 JSON 格式，JSON 格式的数据与集合对象之间的转换是非常简单且有效率的；而且前者逻辑简单，易于使用与维护。

(6) 在 AndroidManifest.xml 文件中对 CustomAdapterActivity 进行注册，编写代码如下：

```
<activity android:name=".custom_adapter.CustomAdapterActivity" />
```

(7) 运行程序，显示效果如图 4-8 所示。分别点击 ListView 的 Item 和 Item 中的头像图片，可以得到不同的信息提示。

需要注意的是，当开发者在 Item 的布局中添加了 Button 等可以自动获取事件的组件时，可能会发现 ListView 的 OnItemClickListener 事件没有响应，这是因为此类控件会“占领”整个 Item 的焦点，导致 Item 焦点被“截取”。解决这种情况的办法是：在 Item 的布局文件最外层的布局控件中(LinearLayout、RelativeLayout 等)添加如下属性，代码如下：

```
android:descendantFocusability="blocksDescendants"
```

该属性的功能是：当 ViewGroup 中的一个 View 控件获取焦点时，定义 ViewGroup 和

其子控件之间的关系，属性值有三种。

✧ beforeDescendants：ViewGroup 会优先其子类控件而获取焦点。
✧ afterDescendants：ViewGroup 只有当其子类控件不需要获取焦点时才获取焦点。
✧ blocksDescendants：ViewGroup 会覆盖其子类控件而直接获得焦点。

4.2.6　自定义 Adapter 的优化

通过继承 BaseAdapter 类创建的 Adapter，用途广泛，通常一个 Android 程序中会存在许多不同的自定义的 Adapter，它通常是为实现更加复杂的列表显示效果与需求来定制的。因此，对于如何优化 Adapter 才能够尽可能地提高用户体验，提高程序的健壮性，是一个非常关键的问题。

对于自定义 Adapter 的优化，通常采用以下三种方法：

(1) 复用布局视图对象：此方法在 4.2.5 小节中已做介绍，主要是在 getView()方法中，对布局视图(convertView 对象)进行非空判断。

(2) 图片异步加载：列表中的图片处理问题往往是非常致命的，因为图片的大小直接影响内存的占用问题，再者，加载网络图片时必然会发生内存溢出等问题。解决方法通常是采用一套成熟的框架或是工具类，而不是开发者自己重新编写代码来解决，因为成熟的框架是经过不断测试与验证得到的。

(3) 组件重用：4.2.5 小节中已经对 getView()方法进行了讲解，细心的读者会发现，虽然对布局视图(convertView 对象)进行了非空判断，但是其中的控件还是会不断创建，造成了程序内存不必要的消耗。解决方法是，可以在自定义的 Adapter 中创建一个内部类，用于存放布局视图中所有控件的对象。

下述示例用于实现：在 4.2.5 小节中创建的“CustomAdapter.java”基础上进行修改，实现组件重用功能。修改代码如下：

```
public class CustomAdapter extends BaseAdapter {

	private Context ctx;
	// 数据源
	private List<Animal> datas = null;

	public CustomAdapter(Context ctx, List<Animal> datas) {
		super();
		this.ctx = ctx;
		this.datas = datas;
	}

	@Override
	public int getCount() {
		return datas.size();
```

```
}

@Override
public Object getItem(int position) {
        return datas.get(position);
}

@Override
public long getItemId(int position) {
        return position;
}

@Override
public View getView(int position, View convertView, ViewGroup parent) {

        // 从数据源中取出当前要显示的对象
        final Animal animal = datas.get(position);
        ViewHolder holder = null;

        if (convertView == null) {
                // 为了避免重复加载资源影响性能，只有在 contentView 为空时才加载
                convertView = LayoutInflater.from(ctx).inflate(
                                R.layout.item_adapter_layout, parent, false);

                // 如果 convertView 是 null，则创建新的 ViewHolder
                holder = new ViewHolder();
                // 引用加载布局文件中的控件
                holder.picIv = (ImageView) convertView
                                .findViewById(R.id.item_pic_iv);
                holder.nameTv = (TextView) convertView
                                .findViewById(R.id.item_name_tv);
                holder.introTv = (TextView) convertView
                                .findViewById(R.id.item_introduce_tv);

                // 将 ViewHolder 对象以“tag”的方式绑定到当前 convertView
                convertView.setTag(holder);
        } else {
                // 如果 convertView 不为 null，则从 convertView 的 tag 中取出
                holder = (ViewHolder) convertView.getTag();
        }
```

```
            // 之后直接对 ViewHolder 中的对象进行操作
            // 将需要显示的信息添加到控件
            holder.picIv.setImageResource(animal.getPic());
            holder.nameTv.setText(animal.getName());
            holder.introTv.setText(animal.getIntroduce());

            // 为头像图片添加点击事件
            holder.picIv.setOnClickListener(new OnClickListener() {

                @Override
                public void onClick(View arg0) {
                    Toast.makeText(ctx, animal.getIntroduce(),
                    Toast.LENGTH_SHORT).show();
                }
            });

            // 返回(提交)当前处理完毕的 Item 布局
            return convertView;
        }

        class ViewHolder {
            ImageView picIv;
            TextView nameTv;
            TextView introTv;
        }
}
```

上述代码中，首先对“convertView”进行非空判断，如果为空，则创建新的 ViewHolder 对象，同时将 ViewHolder 对象中的控件初始化，通过 convertView 的 setTag()方法将 ViewHolder 对象与其绑定；如果不为空，则说明之前已经存在 convertView 对象，此时，只需通过 convertView 的 getTag()方法将与之绑定的 ViewHolder 对象取出即可。之后，只对 ViewHolder 对象中的控件对象进行操作。如此以来，会大大提高列表的加载效率。

4.3 PopupWindow

PopupWindow(弹窗)是一种悬浮在当前 Activity 之上的视图，用于显示任意的 View，也可以灵活地指定其显示的位置。

4.3.1 PopupWindow 概述

显示 PopupWindow，至少需要指定三个条件：内容视图(contentView)、宽度(width)、高度(height)。PopupWindow 的构造方法大约有九个，常用的构造方法如表 4-1 所示。

表 4-1 PopupWindow 的常用构造方法

方 法 名	描 述
PopupWindow (Context context)	创建一个 PopupWindow 对象
PopupWindow(View contentView)	通过内容视图创建一个 PopupWindow 对象
PopupWindow(View contentView, int width, int height)	通过内容视图创建一个 PopupWindow 对象，同时设置大小
PopupWindow(View contentView, int width, int height, boolean focusable)	同上，focusable 表示是否获取焦点，当为 true 时，焦点被 PopupWindow 占用，当前 Activity 无法直接与用户交互，默认为 false

创建 PopupWindow 后，还可以对其进行一些设置，常用方法如表 4-2 所示。

表 4-2 PopupWindow 其他常用设置方法

方 法 名	描 述
setContentView(View contentView)	设置 PopupWindow 的内容视图
getContentView()	获取 PopupWindow 的内容视图
setTouchable(boolean touchable)	设置 PopupWindow 是否响应 touch 事件，如果为 false，作用于 PopupWindow 的所有 touch 事件将无效，默认为 true
setFocusable(boolean focusable)	设置 PopupWindow 是否获取焦点，同构造方法中的“focusable”
setOutsideTouchable(boolean touchable)	设置 PopupWindow 以外的区域是否可以点击，为 true 时，点击以外的区域将自动关闭 PopupWindow，前提是，需要对 PopupWindow 设置背景资源
setBackgroundDrawable(Drawable background)	为 PopupWindow 设置背景资源，除此之外，只有设置了背景资源后，setOutsideTouchable()才能生效，并且点击手机的“返回”按键时，可关闭 PopupWindow(否则可能不会有任何响应)
showAsDropDown(View anchor)	相对于指定控件的左下方显示 PopupWindow
showAsDropDown(View anchor, int xoff, int yoff)	相对于指定控件的左下方显示 PopupWindow，同时设置偏移量
showAtLocation(View parent, int gravity, int x, int y)	相对于父控件的位置显示 PopupWindow，gravity 值可以是 Gravity.CENTER、Gravity.BOTTOM 等，同时设置偏移量
setAnimationStyle(int)	设置动画效果
dismiss()	关闭 PopupWindow

4.3.2 PopupWindow 的使用

下述示例用于实现：点击按钮显示 PopupWindow，点击 PopupWindow 以外的区域可关闭 PopupWindow。PopupWindow 中有两个按钮，单击第一个按钮给出 Toast 消息提示；单击第二个按钮可关闭 PopupWindow。

(1) 创建新的项目，首先为 PopupWindow 创建内容视图布局文件“pop_layout.xml”，代码如下：

```
<?xml version="1.0" encoding="utf-8"?>
<LinearLayout xmlns:android="http://schemas.android.com/apk/res/android"
        android:layout_width="match_parent"
        android:layout_height="match_parent"
        android:background="#ccc"
        android:orientation="vertical" >

    <Button
        android:id="@+id/pop_layout_hello_btn"
        android:layout_width="wrap_content"
        android:layout_height="wrap_content"
        android:padding="5dp"
        android:text="打声招呼"
        android:textSize="18sp" />

    <Button
        android:id="@+id/pop_layout_close_btn"
        android:layout_width="wrap_content"
        android:layout_height="wrap_content"
        android:padding="5dp"
        android:text="关闭弹窗"
        android:textSize="18sp" />

</LinearLayout>
```

上述代码，在布局中添加了两个按钮，分别是“打声招呼”和“关闭弹窗”。

(2) 修改 Activity 的布局文件“activity_main.xml”，代码如下：

```
<RelativeLayout xmlns:android="http://schemas.android.com/apk/res/android"
    xmlns:tools="http://schemas.android.com/tools"
    android:layout_width="match_parent"
    android:layout_height="match_parent"
    android:background="#ededed" >
```

```
    <Button
        android:id="@+id/main_showpop_btn"
        android:layout_width="wrap_content"
        android:layout_height="wrap_content"
        android:text="Show PopupWindow" />

</RelativeLayout>
```

上述代码，在布局中只添加了一个按钮，用于点击后打开 PopupWindow。

(3) 修改“MainActivity.java”，代码如下：

```
public class MainActivity extends Activity {
    private Button showPopBtn;
    private Context ctx;

    @Override
    protected void onCreate(Bundle savedInstanceState) {
        super.onCreate(savedInstanceState);
        setContentView(R.layout.activity_main);

        ctx = this;
        showPopBtn = (Button) findViewById(R.id.main_showpop_btn);
        showPopBtn.setOnClickListener(new View.OnClickListener() {
            @Override
            public void onClick(View view) {
                showPopupWindow(view);
            }
        });
    }

    /**
     * 显示弹窗
     * @param refView 参照控件
     */
    private void showPopupWindow(View refView) {
        // 构造弹窗布局视图
        View popLayout = LayoutInflater.from(ctx)
                .inflate(R.layout.pop_layout,null);

        // 创建 PopupWindow，设置参照控件和大小，参数依次是加载的 View
```

```
        final PopupWindow popWindow = new PopupWindow(popLayout,
                        ViewGroup.LayoutParams.WRAP_CONTENT,
                        ViewGroup.LayoutParams.WRAP_CONTENT, true);

        // 设置 PopupWindow 以外的区域是否可以点击
        popWindow.setOutsideTouchable(true);
        // 为 PopupWindow 设置一个透明背景
        popWindow.setBackgroundDrawable(
                new ColorDrawable(Color.TRANSPARENT));

        // 显示 PopupWindow
        popWindow.showAsDropDown(refView, 10, 0);

        // 引用 PopupWindow 中的两个按钮并添加点击事件
        Button helloBtn = (Button) popLayout
                        .findViewById(R.id.pop_layout_hello_btn);
        Button closeBtn = (Button) popLayout
                        .findViewById(R.id.pop_layout_close_btn);
        helloBtn.setOnClickListener(new View.OnClickListener() {
                @Override
                public void onClick(View v) {
                        Toast.makeText(MainActivity.this,
                         "你好，我是弹窗", Toast.LENGTH_SHORT).show();
                }
        });
        closeBtn.setOnClickListener(new View.OnClickListener() {
                @Override
                public void onClick(View v) {
                        Toast.makeText(MainActivity.this,
                         "关闭了弹窗", Toast.LENGTH_SHORT).show();
                        // 关闭弹窗
                        popWindow.dismiss();
                }
        });
    }
}
```

上述代码中，核心方法是 showPopupWindow(View refView)，参数“refView”是显示 PopupWindow 的参照控件对象，程序运行效果如图 4-9 所示。

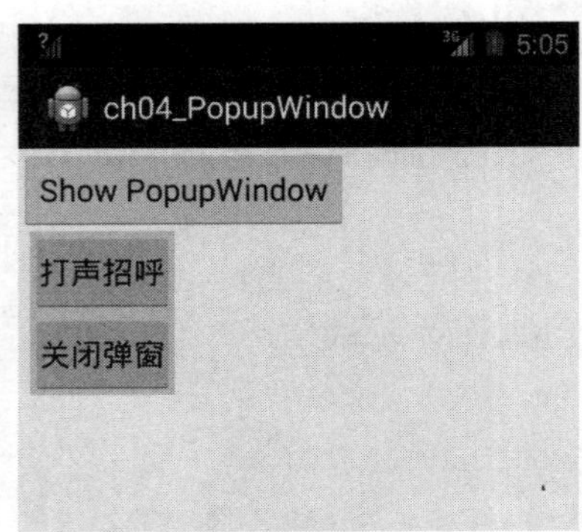

图 4-9　PopupWindow 程序效果

4.4　ViewPager

ViewPager 控件是一个允许用户左右滑动切换页面的布局管理器，继承 ViewGroup 类。通过适配器(PagerAdapter)管理要显示的页面，就像 ListView 控件与数据之间，需要通过适配器进行关联。

ViewPager 控件在 Android3.0 开始时被支持。在创建项目时，会自动导入 JAR 包“android-support-v4.jar”，对于低版本的 Android SDK，需要手动导入该 JAR 包。

4.4.1　ViewPager 概述

在使用 ViewPager 控件之前，首先需要了解 ViewPager 控件相关的类以及使用方法。该控件的使用方法比较简单，有时需要重写该控件来实现更复杂的需求。本小节只介绍 ViewPager 控件基本使用方法。

1. ViewPager 控件

ViewPager 控件基本的功能就是左右滑动切换不同页面，也可以手动设置当前显示的页面，常用方法如表 4-3 所示。

表 4-3　ViewPager 控件常用方法

方 法 名	描 述
addView(View child, int index, ViewGroup.LayoutParams params)	添加子 View。 ✧child：子 View 对象 ✧index：子 View 在 ViewPager 中的位置 ✧params：布局对象
getCurrentItem()	获取当前子 View 所在的索引位置，返回 int 类型值
setAdapter(PagerAdapter adapter)	设置适配器
setCurrentItem(int item)	改变当前显示页面，平滑滚动
setCurrentItem(int item, boolean smoothScroll)	改变当前显示页面，smoothScroll 表示是否平滑滚动

另外，ViewPager 控件还有一个重要的事件监听器 OnPageChangeListener，该监听器用于当监听到 ViewPager 控件页面改变时，发出回调。OnPageChangeListener 监听器的回调方法如表 4-4 所示。

表 4-4　OnPageChangeListener 监听器的回调方法

方 法 名	描　述
onPageScrollStateChanged(int state)	当页面滑动状态改变时。state 表示状态，分为： SCROLL_STATE_IDLE：空闲状态 SCROLL_STATE_DRAGGING：正在滑动状态 SCROLL_STATE_SETTLING：自动沉降状态
onPageScrolled(int position, float positionOffset, int positionOffsetPixels)	当页面正在滑动时。Position 表示正在被滑动的页面索引，其他两个值为偏移量
onPageSelected(int position)	当页面滑动完毕时。position 为当前页面的索引

2. PagerAdapter

PagerAdapter 是 ViewPager 控件使用的适配器，类似于 ListView 控件使用的 BaseAdapter，也是一个抽象类。子类在继承 PagerAdapter 类时，必须重写以下方法：

✧ instantiateItem(ViewGroupcontainer, intposition)：创建要显示的页面并返回。与 BaseAdapter 中的 getView()方法类似。
✧ destroyItem(ViewGroupcontainer, intposition, Objectobject)：从 container 中删除 position 位置的页面。
✧ getCount()：获取要滑动页面的个数。
✧ isViewFromObject(Viewview, Objectobject)：view 与 object 是否为关联对象。

4.4.2　编写简易图片查看器

本小节将通过一个简易图片查看器程序，讲解 ViewPager 控件和 PagerAdapter 适配器的基本使用方法。

下述示例用于实现：编写一个简易图片查看器，要求如下：

✧ 可以通过左右滑动方式切换图片。
✧ 通过按钮控制图片的切换。
✧ 可查看指定页码的图片。

编写简易图片查看器的步骤如下：

(1) 创建项目“ch04_ImageViewer”，将程序使用到的图片放入“res/drawable-hdpi”目录下。修改“activity_main.xml”布局文件，代码如下：

```
<RelativeLayout xmlns:android="http://schemas.android.com/apk/res/android"
    xmlns:tools="http://schemas.android.com/tools"
    android:layout_width="match_parent"
    android:layout_height="match_parent" >

    <android.support.v4.view.ViewPager
```

```
    android:id="@+id/act_viewpager"
    android:layout_width="match_parent"
    android:layout_height="match_parent"
    android:layout_above="@+id/ll" />

<LinearLayout
    android:id="@+id/ll"
    android:layout_width="match_parent"
    android:layout_height="wrap_content"
    android:layout_alignParentBottom="true"
    android:orientation="horizontal" >

<Button
    android:id="@+id/act_main_pre_btn"
    android:layout_width="match_parent"
    android:layout_height="wrap_content"
    android:layout_weight="1"
    android:text="上一张" />

<EditText
    android:id="@+id/act_main_idx_et"
    android:layout_width="match_parent"
    android:layout_height="wrap_content"
    android:layout_weight="1"
    android:inputType="number"
    android:singleLine="true" />

<Button
    android:id="@+id/act_main_jump_btn"
    android:layout_width="match_parent"
    android:layout_height="wrap_content"
    android:layout_weight="1"
    android:text="跳转" />

<Button
    android:id="@+id/act_main_next_btn"
    android:layout_width="match_parent"
    android:layout_height="wrap_content"
    android:layout_weight="1"
    android:text="下一张" />
```

```
    </LinearLayout>
</RelativeLayout>
```

在布局文件中添加了一个 ViewPager 控件，页面下方添加了三个控制按钮和一个输入框，用于控制 ViewPager 控件页面的切换操作。

(2) 修改“MainActivity.java”类，简单起见，ViewPager 控件的适配器类以内部类的形式编写到“MainActivity.java”类中，基本代码如下：

```
public class MainActivity extends Activity {
    /** 图片资源 ID */
    private static final int[] picIds = { R.drawable.pic1,
            R.drawable.pic2,R.drawable.pic3, R.drawable.pic4,
            R.drawable.pic5 };
    private Button preBtn = null;
    private Button nextBtn = null;
    private Button jumpBtn = null;
    private EditText idxEt = null;
    private ViewPager viewPager = null;
    // 当前页面索引
    private int currentIdx = 0;

    @Override
    protected void onCreate(Bundle savedInstanceState) {
        super.onCreate(savedInstanceState);
        setContentView(R.layout.activity_main);

        preBtn = (Button) findViewById(R.id.act_main_pre_btn);
        nextBtn = (Button) findViewById(R.id.act_main_next_btn);
        jumpBtn = (Button) findViewById(R.id.act_main_jump_btn);
        idxEt = (EditText) findViewById(R.id.act_main_idx_et);
        viewPager = (ViewPager) findViewById(R.id.act_viewpager);

        preBtn.setOnClickListener(onBtnClickListener);
        nextBtn.setOnClickListener(onBtnClickListener);
        jumpBtn.setOnClickListener(onBtnClickListener);
        viewPager.setAdapter(new PicPagerAdapter());
        // 添加页面改变监听器
        viewPager.addOnPageChangeListener(new OnPageChangeListener() {

            @Override
            public void onPageSelected(int position) {
                currentIdx = position;
```

```
                }

                @Override
                public void onPageScrolled(int position,
                        float positionOffset,int positionOffsetPixels) {
                }
                @Override
                public void onPageScrollStateChanged(int state) {

                }
            });
        }

        private OnClickListener onBtnClickListener = new OnClickListener() {

            @Override
            public void onClick(View v) {
                if (v == preBtn) {
                    // 上一张
                    int idx = currentIdx - 1;
                    if (idx >= 0) {
                        viewPager.setCurrentItem(idx);
                    }
                } else if (v == nextBtn) {
                    // 下一张
                    int idx = currentIdx + 1;
                    if (idx < picIds.length) {
                        viewPager.setCurrentItem(idx);
                    }
                } else if (v == jumpBtn) {
                    // 跳转到指定页码
                    // 获取用户输入的页码
                    String idxStr = idxEt.getText().toString();
                    if (!idxStr.isEmpty()) {
                        // 因为下标从 0 开始，页码应-1
                        int idx = Integer.parseInt(idxEt.getText()
                                .toString()) - 1;
                        if (idx >= 0 && idx < picIds.length)
                        {
                            viewPager.setCurrentItem(idx);
```

```
                        }
                    }
                }
            }
        };
}
```

上述代码中，主要通过 addOnPageChangeListener()方法对 ViewPager 控件添加了 OnPageChangeListener 监听器，当页面改变后，记录当前页面的索引值；为三个控制按钮分别添加了 OnClickListener 监听器，用于控制页面的跳转。

接下来继续在“MainActivity.java”类中，为 ViewPager 控件创建适配器内部类，代码如下：

```
public class PicPagerAdapter extends PagerAdapter {

    @Override
    public int getCount() {
        return picIds.length;
    }

    @Override
    public boolean isViewFromObject(View view, Object object) {
        return view == object;
    }

    @Override
    public void destroyItem(ViewGroup container, int position,
        Object object) {
        View view = (View) object;
        container.removeView(view);
    }

    @Override
    public Object instantiateItem(ViewGroup container, int position) {
        ImageView imageView = new ImageView(MainActivity.this);
        imageView.setImageResource(picIds[position]);
        container.addView(imageView);
        return imageView;
    }
}
```

运行程序，效果如图 4-10 所示。

图 4-10　图片查看器运行界面

本 章 小 结

(1) Draw 9-patch 工具用于对普通图片进行处理，以适用于 Android UI 显示，使之不会因屏幕分辨率不同而出现失真现象。

(2) Draw 9-patch 工具可以控制图片的拉伸区域和内容填充区域的自动缩放。

(3) ListView 绑定的适配器通常有 ArrayAdapter、SimpleAdapter，以及自定义 Adapter。

(4) 自定义的 Adapter 使用起来更加灵活，可实现自定义的功能，但其开发相对复杂。

(5) ArrayAdapter、SimpleAdapter 以及自定义 Adapter，不仅只适用于 ListView 控件，同样也适用于 GridView 和 Spinner 等控件。

本 章 练 习

(1) Draw 9-patch 工具需要自行下载安装后才可以使用。

(A) 对　　(B) 错

(2) 可以使用自定义适配器实现大多数复杂的列表实现。

(A) 对　　(B) 错

(3) ArrayAdapter、SimpleAdapter 以及自定义 Adapter 不能用于下列哪个控件？

(A) ListView　　(B) GridView　　(C) ImageView　　(D) Spinner

(4) 使用 Draw 9-patch 工具，将程序中所需图片进行修改，使之能够适应各种分辨率屏幕。

(5) 使用 ViewPager 控件，编写新的图片查看器，要求可动态添加或删除页面。

第 5 章　传感器

本章目标

- 了解手机中支持的传感器
- 了解手机中各传感器的用途
- 理解常用传感器编程

传感器(Sensor)是一种检测装置，能感受到被测量的信息，并能将感受到的信息，按一定规律变换成电信号或其他所需形式的信息输出，以满足信息的传输、处理、存储、显示、记录和控制等要求。它是实现自动检测和自动控制的首要环节。传感器的存在和发展，让物体有了触觉、味觉和嗅觉等感官，让物体慢慢变得活了起来。本章主要讲解手机中各种传感器的使用。

5.1 传感器简介

智能手机的出现，使得传感器在人们的日常生活中得到了更加广泛的应用，例如指南针，测量海拔、气压工具，“摇一摇”，翻转静音，以及拨打电话时屏幕自动熄灭等，这些功能都是运用各种传感器来实现的。

5.1.1 传感器相关类

1. Sensor

Android 中通过 Sensor 类对每个传感器做了抽象处理，其中包含了一个常量集合，用于描述当前 Sensor 对象所表示传感器的类型，这些传感器大致分为两种类型：原始传感器和虚拟(复合)传感器。原始传感器是指设备中实际存在的物理设备，给出从传感器获得的原始数据；虚拟传感器是指通过多个原始传感器相结合，通过修改或组合原始数据来获得更容易被使用的复合数据，例如手机中的指南针程序，给出了方向的值以及加速计提供的倾斜度等。

Android4.0 中支持的传感器类型如表 5-1 所示。当然，至于这些传感器 API 能否实现，取决于具体的手机是否配有相应的传感器。

表 5-1 Android4.0 支持的传感器类型

类 型 常 量	功 能 描 述	分 类
Sensor.TYPE_LIGHT	光线传感器	原始传感器
Sensor.TYPE_PROXIMITY	距离传感器	
Sensor.TYPE_PRESSURE	气压传感器	
Sensor.TYPE_TEMPERATURE	温度传感器	
Sensor.TYPE_ACCELEROMETER	加速度传感器	
Sensor.TYPE_GYROSCOPE	陀螺仪传感器	
Sensor.TYPE_MAGNETIC_FIELD	磁场传感器	
Sensor.TYPE_RELATIVE_HUMIDITY	相对湿度传感器	
Sensor.TYPE_AMBIENT_TEMPERATURE	环境温度传感器	
Sensor.TYPE_ROTATION_VECTOR	旋转矢量传感器	虚拟(复合)传感器
Sensor.TYPE_LINEAR_ACCELERATION	线性加速度传感器	
Sensor.TYPE_GRAVITY	重力传感器	
Sensor.TYPE_ORIENTATION	方向传感器	

Sensor 类中还描述了传感器的属性，例如：类型、名称、制造商、精确度和范围等。Sensor 类常用方法如表 5-2 所示。

表 5-2　Sensor 类常用方法

方　法　名	描　　述
getMaximumRange()	最大范围
getMinDelay ()	最小延迟
getName ()	名称
getPower ()	功率
getResolution ()	分辨率
getType ()	类型
getVendor ()	供应商
getVersion ()	版本号

2．SensorManager

通过 SensorManager，既可以访问设备的传感器，也可用于获取传感器、添加监听器等操作，常用方法如表 5-3 所示。

表 5-3　SensorManager 类常用方法

方　法　名	描　　述
getDefaultSensor()	根据传入的传感器类型，获取默认传感器
getSensorList()	根据传入的传感器类型，获取所有匹配的传感器
registerListener()	为传感器注册监听事件
unregisterListener()	取消注册监听事件

3．SensorEventListener

SensorEventListener 是传感器事件监听器接口。当程序需要监听传感器的相应事件时，就可以创建一个实现了 SensorEventListener 接口的类，并将这个类注册到 SensorManager 中。SensorEventListener 接口包含两个回调方法：

(1) onSensorChanged(SensorEvent event)：监控传感器值的变化，该方法中的参数是一个 SensorEvent 对象，该对象包含了传感器输出到程序的信息，主要包含四个描述传感器事件的属性。

- ✧ sensor：传感器对象。
- ✧ timestamp：事件发生的时间，单位为毫秒。
- ✧ values：传感器的数据，float 类型的数组。
- ✧ accuracy：传感器输出的精确度分为四个等级，在此，精确度指的是输出值的可靠程度而不是物理值的接近程度。该属性可以有以下四个值。
  - ➢ SensorManager.SENSOR_STATUS_ACCURACY_HIGH：传感器的精确度是最高精确度，数据可靠性高。
  - ➢ SensorManager.SENSOR_STATUS_ACCURACY_LOW：传感器的精确度很低并需要校准，数据可靠性低。
  - ➢ SensorManager.SENSOR_STATUS_ACCURACY_MEDIUM：传感器的数据

具有平均精确度，校准可能会改善效果，数据可靠性一般。

➢ SensorManager.SENSOR_STATUS_UNRELIABLE：传感器数据不可靠，需要校准且当前不能读取数据。

(2) onAccuracyChanged(Sensor sensor, int accuracy)：监控传感器精确度的变化。

5.1.2 查看本机传感器

因为不是所有手机都配有相同传感器，因此在使用传感器之前，首先要了解手机中是否存在所需传感器。查看方式主要有三种：阅读手机说明书、网上查询或编写代码来查询。

下述示例用于实现：通过编写代码的方式查看当前手机所配有的传感器列表。

(1) 创建新项目(ch05_FindSensor)，编写布局文件“activity_main.xml”，代码如下：

```
<ScrollView xmlns:android="http://schemas.android.com/apk/res/android"
        android:layout_width="match_parent"
        android:layout_height="match_parent"
        android:orientation="horizontal" >

<TextView
        android:id="@+id/act_main_content_tv"
        android:layout_width="match_parent"
        android:layout_height="wrap_content" />

</ScrollView>
```

(2) 编写“MainActivity.java”类，代码如下：

```
public class MainActivity extends Activity {
        private TextView contentTv;
        // 传感器管理器
        private SensorManager sensorManager;

        @Override
        protected void onCreate(Bundle savedInstanceState) {
                super.onCreate(savedInstanceState);
                setContentView(R.layout.activity_main);

                contentTv = (TextView) findViewById(R.id.act_main_content_tv);

                // 获取传感器管理器
                sensorManager =
                        (SensorManager) getSystemService(Context.SENSOR_SERVICE);
```

```
// 获取当前手机所支持的所有传感器
List<Sensor> allSensors =
        sensorManager.getSensorList(Sensor.TYPE_ALL);

StringBuffer sbf = new StringBuffer();
sbf.append("当前手机共有" + allSensors.size() + "个传感器：\n");

for (Sensor sensor : allSensors) {
        switch (sensor.getType()) {
        case Sensor.TYPE_LIGHT:
                sbf.append(sensor.getType()
                 + " 光线传感器(Light sensor) \n");
                break;
        case Sensor.TYPE_PROXIMITY:
                sbf.append(sensor.getType()
                 + " 距离传感器(Proximity sensor) \n");
                break;
        case Sensor.TYPE_PRESSURE:
                sbf.append(sensor.getType()
                 + " 气压传感器(Pressure sensor) \n");
                break;
        case Sensor.TYPE_TEMPERATURE:
                sbf.append(sensor.getType()
                 + " 温度传感器(Temperature sensor) \n");
                break;
        case Sensor.TYPE_ACCELEROMETER:
                sbf.append(sensor.getType()
                + " 加速度传感器(Accelerometer sensor) \n");
                break;
        case Sensor.TYPE_GYROSCOPE:
                sbf.append(sensor.getType()
                 + " 陀螺仪传感器(Gyroscope sensor) \n");
                break;
        case Sensor.TYPE_MAGNETIC_FIELD:
                sbf.append(sensor.getType()
                + " 磁场传感器(Magnetic field sensor) \n");
                break;
        case Sensor.TYPE_RELATIVE_HUMIDITY:
                sbf.append(sensor.getType()
                + " 相对湿度传感器(Relative humidity sensor) \n");
```

```
                    break;
                case Sensor.TYPE_AMBIENT_TEMPERATURE:
                    sbf.append(sensor.getType()
                    + " 环境温度传感器(Ambient temperature sensor) \n");
                    break;
                case Sensor.TYPE_ROTATION_VECTOR:
                    sbf.append(sensor.getType()
                    + " 旋转矢量传感器(Rotation vector sensor) \n");
                    break;
                case Sensor.TYPE_LINEAR_ACCELERATION:
                    sbf.append(sensor.getType()
                    + " 线性加速度传感器(Linear acceleration sensor) \n");
                    break;
                case Sensor.TYPE_GRAVITY:
                    sbf.append(sensor.getType()
                     + " 重力传感器(Gravity sensor) \n");
                    break;
                case Sensor.TYPE_ORIENTATION:
                    sbf.append(sensor.getType()
                     + " 方向传感器(Orientation sensor) \n");
                    break;
                default:
                    sbf.append(sensor.getType() + " 其他传感器 \n");
                    break;
                }
                sbf.append("设备供应商： " + sensor.getVendor()
                      + "\n 设备版本号："+ sensor.getVersion() + "\n--\n");
            }
            contentTv.setText(sbf.toString());
        }
}
```

上述代码主要实现了查看当前手机所配备的所有传感器，根据手机的不同，传感器的数量也会有所不同，可能会存在一部分无法识别的传感器(其他传感器)，这一部分传感器通常是手机厂商配备的特有的传感器，这里不需要关注。

注 意　在测试传感器程序时，建议尽量使用真机测试，因为模拟器所带的传感器非常少，而且不方便测试，本章中用到的真机为小米手机 2。

(3) 程序运行结果如图 5-1 所示。

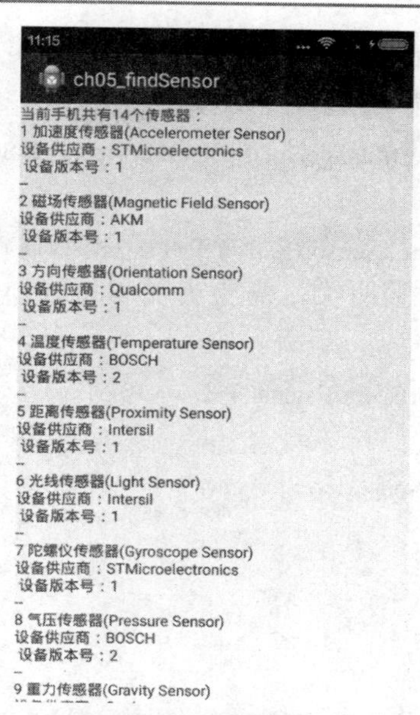

图 5-1　查看本机传感器

5.2　传感器的应用

1．空间直角坐标系的概念

Android 中，传感器通常是使用一个标准的三维空间直角坐标系来表示值、方向，或状态。如图 5-2 所示，空间直角坐标系分为 X 轴、Y 轴和 Z 轴，三条轴相交的位置为原点，正方向为正数，反方向为负数。

✧ X 轴：屏幕水平方向(横向)，向右为正方向。
✧ Y 轴：屏幕垂直方向(纵向)，向上为正方向。
✧ Z 轴：垂直于屏幕，屏幕正前方为正方向。

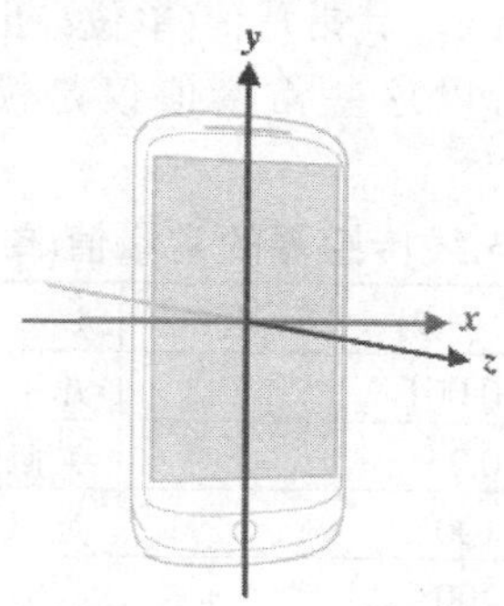

图 5-2　空间直角坐标系图

2．使用传感器的步骤

所有传感器的使用步骤大致相同，通常不会直接获取传感器上面的信息，更多的是获取传感器采集到的数据，比如加速度值、大气压等，具体步骤如下所示：

(1) 获得 SensorManager 传感器管理器对象。

```
SensorManager sensorManager= (SensorManager)getSystemService(SENSOR_SERVICE);
```

(2) 获取所需传感器对象。

```
Sensor sensor = sensorManager.getDefaultSensor(Sensor.TYPE_GRAVITY);
```

(3) 创建自定义传感器事件监听类，需要实现 SensorEventListener 接口，根据具体需求编写代码。

```
class SensorListener implements SensorEventListener {
    @Override
    public void onSensorChanged(SensorEvent event) {


    }


    @Override
    public void onAccuracyChanged(Sensor sensor, int accuracy) {


    }
};
```

(4) 根据需要，调用 SensorManager 中相应的 registerListener()方法来注册监听器。

(5) 程序关闭之前或不再使用传感器时，调用 SensorManager 的 unregisterListener()方法来取消注册。

5.2.1 光线传感器

光线传感器(Sensor.TYPE_LIGHT)通常隐藏于手机听筒附近的圆孔内，是一个可以感知光线强度的光电二极管，当光线射入时产生电压，通常被用于根据手机所处环境的光线来自动调节手机屏幕的亮度，从而达到省电、保护视力的作用。

光线传感器输出值的单位为 lux，其常量值(单位：lux)如表 5-4 所示。因为无法准确地用常用语言来做定性描述，所以这些常量值仅是模糊的定义。这些常量被定义在 SensorManager 类中。

表 5-4 光线传感器的常量值(单位：lux)

常 量 名	对 应 值	描 述
LIGHT_NO_MOON	0.001	几乎无光的环境
LIGHT_FULLMOON	0.25	类似于夜间足够的月光环境
LIGHT_CLOUDY	100	类似于乌云密布时的亮度
LIGHT_SUNRISE	400	类似于日出时的亮度
LIGHT_OVERCAST	10000	类似于阴天时的亮度
LIGHT_SHADE	20000	类似于日光条件下，非阳光直射时的亮度
LIGHT_SUNLIGHT	110000	类似于阳光直射的亮度
LIGHT_SUNLIGHT_MAX	120000	类似于阳光直射时最大的亮度

下述代码用于实现：通过编写程序实现光线传感器示例观察手机在不同的光照场景下，数值的变化。

(1) 创建新项目(ch05_LightSensor)，编写布局文件“activity_main.xml”，代码如下：

```
<RelativeLayout xmlns:android="http://schemas.android.com/apk/res/android"
    xmlns:tools="http://schemas.android.com/tools"
    android:layout_width="match_parent"
    android:layout_height="match_parent"
    android:background="#ededed"
    android:padding="10dp" >

    <TextView
        android:id="@+id/act_main_content_tv"
        android:layout_width="wrap_content"
        android:layout_height="wrap_content"
        android:layout_marginTop="10dp"
        android:text="光线强度值:"
        android:textSize="20sp" />

</RelativeLayout>
```

(2) 编写“MainActivity.java”类，代码如下：

```
public class MainActivity extends Activity {

        private TextView contentTv;
        /** 传感器管理器 */
        private SensorManager sensorManager;
        /** 光线传感器 */
        private Sensor lightSensor;

        private SensorEventListener sensorListener;

        @Override
        protected void onCreate(Bundle savedInstanceState) {
                super.onCreate(savedInstanceState);
                setContentView(R.layout.activity_main);

                contentTv = (TextView) findViewById(R.id.act_main_content_tv);

                sensorManager = (SensorManager) getSystemService(SENSOR_SERVICE);
                lightSensor = sensorManager.getDefaultSensor(Sensor.TYPE_LIGHT);
                // 实例化传感器事件监听器
```

```
            sensorListener = new SensorListener();
            sensorManager.registerListener(sensorListener, lightSensor,
                        SensorManager.SENSOR_DELAY_UI);
    }

    class SensorListener implements SensorEventListener {

            @Override
            public void onSensorChanged(SensorEvent event) {
                    contentTv.setText("光线强度值：" + event.values[0]);
            }

            @Override
            public void onAccuracyChanged(Sensor sensor, int accuracy) {

            }
    };

    @Override
    protected void onDestroy() {
            super.onDestroy();
            // 取消监听
            sensorManager.unregisterListener(sensorListener);
    }
}
```

光线传感器事件监听器中的 onSensorChanged()回调方法中，“event.values”数组中只有一个值，用来表示光照强度，程序运行效果如图 5-3 所示。

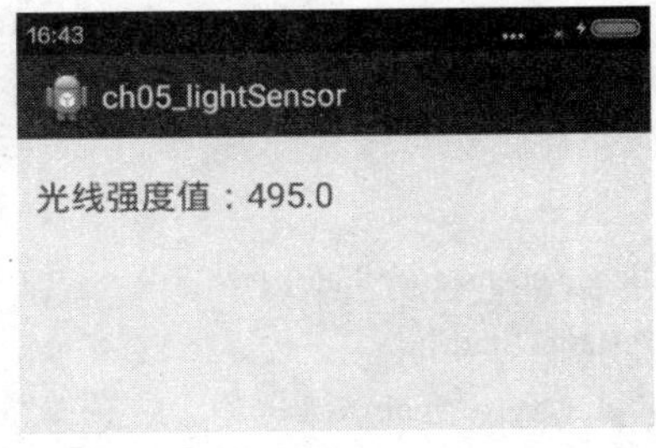

图 5-3　光线传感器

5.2.2　距离传感器

距离传感器(Sensor.TYPE_PROXIMITY)又称接近传感器，该设备也是隐藏于手机听筒

附近的圆孔内，是一个弱红外 LED(发光二极管)。当有物体距离该传感器足够近时，光电探测器会检测到反射的红外光。距离传感器在手机中通常用于：拨打电话时，当耳朵贴近听筒，屏幕会自动关闭，以节省电量。

部分距离传感器输出的数据是与物体之间的距离，以 cm 为单位；另一部分用来测量物体是否在一个阈值范围内，通常这部分距离传感器被应用到手机中。距离传感器的动态范围通常在 5 cm 左右，可以通过传感器对象的“getMaximumRange()”方法获取这个值。

下述代码用于实现：通过编写程序实现距离传感器的使用，分别测试当距离传感器有物体遮挡或没有物体遮挡时所显示的数值，并测试距离传感器的有效感应距离。

(1) 创建新项目(ch05_ProxiSensor)，编写布局文件“activity_main.xml”，代码如下：

```
<RelativeLayout xmlns:android="http://schemas.android.com/apk/res/android"
    xmlns:tools="http://schemas.android.com/tools"
    android:layout_width="match_parent"
    android:layout_height="match_parent"
    android:background="#ededed"
    android:padding="10dp" >

<TextView
    android:id="@+id/act_main_content_tv"
    android:layout_width="wrap_content"
    android:layout_height="wrap_content"
    android:layout_marginTop="10dp"
    android:textSize="20sp" />

</RelativeLayout>
```

(2) 编写“MainActivity.java”类，代码如下：

```
public class MainActivity extends Activity {
        private TextView contentTv;
        /** 传感器管理器 */
        private SensorManager sensorManager;
        /** 距离传感器 */
        private Sensor proximitySensor;

        private SensorEventListener sensorListener;

        @Override
        protected void onCreate(Bundle savedInstanceState) {
                super.onCreate(savedInstanceState);
                setContentView(R.layout.activity_main);
```

```
        contentTv = (TextView) findViewById(R.id.act_main_content_tv);

        sensorManager = (SensorManager) getSystemService(SENSOR_SERVICE);
        proximitySensor =
                sensorManager.getDefaultSensor(Sensor.TYPE_PROXIMITY);
        // 实例化传感器事件监听器
        sensorListener = new SensorListener();
        sensorManager.registerListener(sensorListener, proximitySensor,
                        SensorManager.SENSOR_DELAY_UI);
    }

    class SensorListener implements SensorEventListener {

        @Override
        public void onSensorChanged(SensorEvent event) {
            // 保留两位小数
            float distance = Math.round(event.values[0] * 100) / 100;
            contentTv.setText("当前值： " + distance);
            if (distance == 0) {
                contentTv.append("\n\n 靠近传感器");
            } else {
                contentTv.append("\n\n 远离传感器");
            }

        }

        @Override
        public void onAccuracyChanged(Sensor sensor, int accuracy) {

        }
    };

    @Override
    protected void onDestroy() {
        super.onDestroy();
        // 取消监听
        sensorManager.unregisterListener(sensorListener);
    }
}
```

上述代码中，对距离传感器进行事件监听。运行后，屏幕会显示距离传感器当前检测

到的值，当被遮挡时，显示“靠近传感器”；没有被遮挡时，显示“远离传感器”。经测试(所测试的手机型号是小米 2)，距离值只有两个：0 和 5。没有被遮挡时，值为 5；被遮挡时，值为 0。当物体靠近传感器大约 2 厘米时，值会从 5 变为 0，这说明传感器自认为被遮挡了。当然，具体的参数会根据手机的不同而不同，但被遮挡时，通常值都为 0。运行效果如图 5-4 所示。

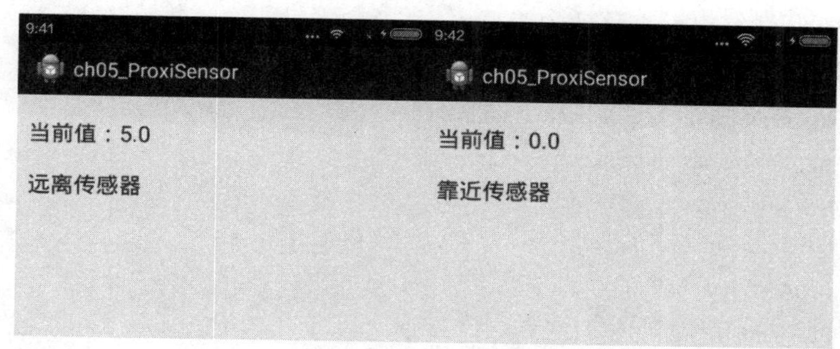

图 5-4　距离传感器

5.2.3　气压传感器

气压传感器(Sensor.TYPE_PRESSURE)通常用于测量大气气压，通过气压值可以计算出当前海拔高度，配备该传感器的手机设备并不多。气压传感器就像已知压力的腔室之上的一个鼓面一样，随着外部压力的改变，鼓面也随之凸起或凹陷。通常，气压传感器采集到的数据可能会有误差，因为这受空气密度、环境温度等因素的影响。

可以通过调用 SensorManager 类中的 getAltitude(float p0, float p)方法，通过气压计算出海拔高度。该方法中，第一个参数“p0”表示海平面的气压值，第二个参数“p”表示当前测量的气压值，返回值是以米为单位的海拔高度。其中第一个参数“p0”可以是以下两种值：

✧ SensorManager.PRESSURE_STANDARD_ATMOSPHERE
  该常量给出了标准气压，该标准气压是基于相对海拔高度，而不是绝对高度。
✧ 有效或平均海平面压力值
  该值通常可以从机场或气象局获取，该值为相对高度和绝对高度都提供了最好的测量结果。

在这两种值中，通常会采用 PRESSURE_STANDARD_ATMOSPHERE 常量值，除非在需要高精准度的情况下才会考虑使用后者。

SensorManager. getAltitude(float p0, float p)方法中，用到的公式如下：

$$\frac{T_0}{L}\left(1-\left(\frac{p}{p_0}\right)^{\frac{RL}{gM}}\right)=44330*\left(1-\left(\frac{p}{p_0}\right)^{\frac{1}{5.255}}\right)$$

上述公式中，计算结果为海拔高度，T_0 为海平面标准温度，L 为温度递减速率，R 为通用气体常数，g 为重力加速度，M 为干燥空气的摩尔质量，等号右侧的公式已经给出了这些参数的具体数值。从上述公式常量中不难发现，造成气压传感器采集到的数据有所误差的因素和许多环境有关。

通过气压传感器可以计算出相同环境下物体的高度。下述代码用于实现：通过编写程序实现气压传感器的使用；根据已获取的当前气压值，计算当前海拔高度；垂直移动手机，观察高度差，计算出垂直移动的距离。

(1) 创建新项目(ch05_PressSensor)，编写布局文件“activity_main.xml”，代码如下：

```
<RelativeLayout xmlns:android="http://schemas.android.com/apk/res/android"
    xmlns:tools="http://schemas.android.com/tools"
    android:layout_width="match_parent"
    android:layout_height="match_parent"
    android:background="#ededed"
    android:padding="10dp" >

<TextView
        android:id="@+id/act_main_content_tv"
        android:layout_width="wrap_content"
        android:layout_height="wrap_content"
        android:layout_marginTop="10dp"
        android:textSize="20sp" />

</RelativeLayout>
```

(2) 编写“MainActivity.java”类，代码如下：

```
public class MainActivity extends Activity {
        private TextView contentTv;
        /** 传感器管理器 */
        private SensorManager sensorManager;
        /** 气压传感器 */
        private Sensor pressureSensor;

        private SensorEventListener sensorListener;

        @Override
        protected void onCreate(Bundle savedInstanceState) {
                super.onCreate(savedInstanceState);
                setContentView(R.layout.activity_main);

                contentTv = (TextView) findViewById(R.id.act_main_content_tv);
```

```
        sensorManager = (SensorManager) getSystemService(SENSOR_SERVICE);
        pressureSensor =
                sensorManager.getDefaultSensor(Sensor.TYPE_PRESSURE);
        // 实例化传感器事件监听器
        sensorListener = new SensorListener();
        sensorManager.registerListener(sensorListener, pressureSensor,
                        SensorManager.SENSOR_DELAY_UI);
    }

    class SensorListener implements SensorEventListener {

        @Override
        public void onSensorChanged(SensorEvent event) {

            float crtP = event.values[0];
            contentTv.setText("当前气压： " + crtP);
            // 计算海拔 PRESSURE_STANDARD_ATMOSPHERE 是海平面的平均气压
            double height = SensorManager.getAltitude(
                    SensorManager.PRESSURE_STANDARD_ATMOSPHERE, crtP);
            DecimalFormat df = new DecimalFormat("0.00");
            contentTv.append("\n\n 当前海拔： " + df.format(height));
        }

        @Override
        public void onAccuracyChanged(Sensor sensor, int accuracy) {

        }
    };

    @Override
    protected void onDestroy() {
        super.onDestroy();
        // 取消监听
        sensorManager.unregisterListener(sensorListener);
    }
}
```

上述代码中，首先在传感器监听事件中获取了当前气压值(event.values[0])，然后通过SensorManager.getAltitude()方法计算出当前海拔，海拔高度可能不够精准(可能为负数)，因为这和很多环境因素有关，通常需要人为矫正。运行项目进行验证，如图 5-5 所示。

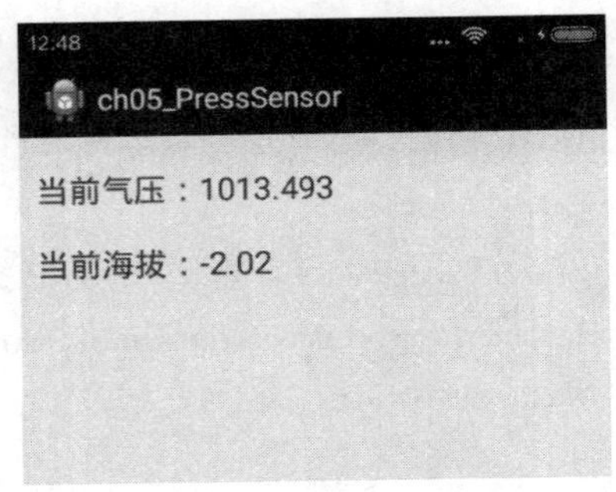

图 5-5　气压传感器

5.2.4　温度传感器

Android 中的温度传感器(Sensor.TYPE_TEMPERATURE)是用于监测 CUP 温度的，并不是用于测量环境温度，因此，实际使用场景并不多见，在 Android 4.0 版本中已经被淘汰，取而代之的是环境温度传感器(Sensor.TYPE_AMBIENT_TEMPERATURE)。对温度传感器感兴趣的读者可以编程进行验证，本章不再讲述。

5.2.5　加速度传感器

加速度传感器(Sensor.TYPE_ACCELEROMETER)在 Android 中是比较常用的传感器之一，主要用于监测手机的运动状态，测量运动的加速度。该传感器有三个数据值，分别表示空间直角坐标系中 X 轴、Y 轴、Z 轴方向上的加速度值，单位为 m/s^2。

下述代码用于实现：通过编写程序实现加速度传感器中 X 轴、Y 轴、Z 轴所对应的加速度值的显示，要求程序界面分为“实时信息”与“最大值信息”两部分，“实时信息”部分用于显示当前加速度值；“最大值信息”用于显示程序运行期间最大加速度的值。

(1) 创建新项目(ch05_AccelSensor)，编写布局文件“activity_main.xml”，代码如下：

```
<LinearLayout xmlns:android="http://schemas.android.com/apk/res/android"
    xmlns:tools="http://schemas.android.com/tools"
    android:layout_width="match_parent"
    android:layout_height="match_parent"
    android:background="#ededed"
    android:orientation="vertical"
    android:padding="10dp" >

<TextView
    android:layout_width="wrap_content"
    android:layout_height="wrap_content"
    android:text="实时信息"
    android:textSize="20sp" />
```

```
<TextView
    android:id="@+id/act_main_info_tv"
    android:layout_width="wrap_content"
    android:layout_height="wrap_content"
    android:layout_marginLeft="20dp"
    android:layout_marginTop="10dp" />

<TextView
    android:layout_width="wrap_content"
    android:layout_height="wrap_content"
    android:layout_marginTop="20dp"
    android:text="最大值信息"
    android:textSize="20sp" />

<TextView
    android:id="@+id/act_main_maxinfo_tv"
    android:layout_width="wrap_content"
    android:layout_height="wrap_content"
    android:layout_marginLeft="20dp"
    android:layout_marginTop="10dp" />

</LinearLayout>
```

(2) 编写“MainActivity.java”类，代码如下：

```
public class MainActivity extends Activity {
        private TextView infoTv = null;
        private TextView maxInfoTv = null;

        /** 传感器管理器 */
        private SensorManager sensorManager;
        /** 加速度传感器 */
        private Sensor accelerometerSensor;
        private SensorEventListener sensorListener;

        /** X 轴最大值 */
        private float maxX = 0;
        /** Y 轴最大值 */
        private float maxY = 0;
        /** Z 轴最大值 */
        private float maxZ = 0;
```

```
    @Override
    protected void onCreate(Bundle savedInstanceState) {
        super.onCreate(savedInstanceState);
        setContentView(R.layout.activity_main);

        infoTv = (TextView) findViewById(R.id.act_main_info_tv);
        maxInfoTv = (TextView) findViewById(R.id.act_main_maxinfo_tv);

        sensorManager = (SensorManager) getSystemService(SENSOR_SERVICE);
        accelerometerSensor = sensorManager
                .getDefaultSensor(Sensor.TYPE_ACCELEROMETER);
        // 实例化传感器事件监听器
        sensorListener = new SensorListener();
        sensorManager.registerListener(sensorListener,
                accelerometerSensor,SensorManager.SENSOR_DELAY_UI);
    }

    class SensorListener implements SensorEventListener {

        @Override
        public void onSensorChanged(SensorEvent event) {
            float[] value = event.values;
            // 获取实时的数据并显示
            float x = value[0];
            float y = value[1];
            float z = value[2];
            String info = " x = " + x + "\n y = " + y + "\n z = " + z;
            infoTv.setText(info);

            // 对实时数据与历史最大值信息进行比较，然后进行显示
            if (x > maxX) {
                maxX = x;
            }
            if (y > maxY) {
                maxY = y;
            }
            if (z > maxZ) {
                maxZ = z;
            }
```

```
                String stmaxInfo =
                        " x = " + maxX + "\n y = "
                        + maxY + "\n z = "+ maxZ;
                maxInfoTv.setText(stmaxInfo);
            }

            @Override
            public void onAccuracyChanged(Sensor sensor, int accuracy) {

            }
        };

        @Override
        protected void onDestroy() {
            super.onDestroy();
            // 取消监听
            sensorManager.unregisterListener(sensorListener);
        }
}
```

上述代码中，实现了对加速度传感器的监听，运行程序后，分别在 X 轴、Y 轴、Z 轴上做加速度操作(沿对应的轴方向快速移动手机)，之后会发现“最大值信息”区域显示的信息中，对应的轴加速度值会明显增大。其中，沿 X 轴方向做加速度操作，效果如图 5-6 所示。

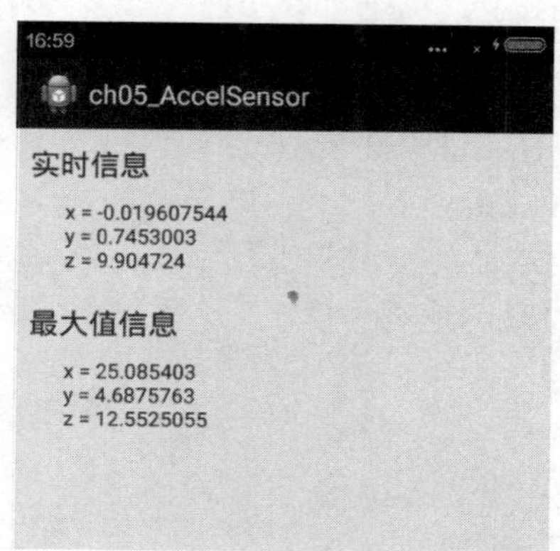

图 5-6　加速度传感器

5.2.6　陀螺仪传感器

陀螺仪传感器(Sensor.TYPE_GYROSCOPE)用于测量设备转动的角速度，与加速度传感器相同的是，该传感器同样有三个数据值，分别是空间直角坐标系中 X 轴、Y 轴、Z 轴方向设备转动的角速度，单位为 m/s^2，原点为设备的左下角。

需要注意的是：使用陀螺仪传感器是不能直接测量角度的，通常可以将陀螺仪的值和时间进行积分来计算角度。陀螺仪传感器的噪声和偏移在计算角度时会导致较大的误差，如果不加以解决，在一定时间内，数据将毫无用处。

下述代码用于实现：与加速度传感器类似，通过编写程序实现陀螺仪传感器中 X 轴、Y 轴、Z 轴所对应的角速度值的显示，要求程序界面分为“实时信息”与“最大值信息”两部分，“实时信息”部分用于显示当前角速度值；“最大值信息”部分用于显示程序运行期间历史最大角速度值。

(1) 创建新项目(ch05_GyrosSensor)，编写布局文件“activity_main.xml”，代码如下：

```
<LinearLayout xmlns:android="http://schemas.android.com/apk/res/android"
    xmlns:tools="http://schemas.android.com/tools"
    android:layout_width="match_parent"
    android:layout_height="match_parent"
    android:background="#ededed"
    android:orientation="vertical"
    android:padding="10dp" >

<TextView
        android:layout_width="wrap_content"
        android:layout_height="wrap_content"
        android:text="实时信息"
        android:textSize="20sp" />

<TextView
        android:id="@+id/act_main_info_tv"
        android:layout_width="wrap_content"
        android:layout_height="wrap_content"
        android:layout_marginLeft="20dp"
        android:layout_marginTop="10dp" />

<TextView
        android:layout_width="wrap_content"
        android:layout_height="wrap_content"
        android:layout_marginTop="20dp"
        android:text="最大值信息"
        android:textSize="20sp" />

<TextView
        android:id="@+id/act_main_maxinfo_tv"
        android:layout_width="wrap_content"
```

```
        android:layout_height="wrap_content"
        android:layout_marginLeft="20dp"
        android:layout_marginTop="10dp" />

</LinearLayout>
```

(2) 编写“MainActivity.java”类，代码如下：

```
public class MainActivity extends Activity {

    private TextView infoTv = null;
    private TextView maxInfoTv = null;

    /** 传感器管理器 */
    private SensorManager sensorManager;
    /** 陀螺仪传感器 */
    private Sensor gyroscopeSensor;
    private SensorEventListener sensorListener;

    /** X 轴最大值 */
    private float maxX = 0;
    /** Y 轴最大值 */
    private float maxY = 0;
    /** Z 轴最大值 */
    private float maxZ = 0;

    @Override
    protected void onCreate(Bundle savedInstanceState) {
        super.onCreate(savedInstanceState);
        setContentView(R.layout.activity_main);

        infoTv = (TextView) findViewById(R.id.act_main_info_tv);
        maxInfoTv = (TextView) findViewById(R.id.act_main_maxinfo_tv);

        sensorManager = (SensorManager) getSystemService(SENSOR_SERVICE);
        gyroscopeSensor = sensorManager
                .getDefaultSensor(Sensor.TYPE_GYROSCOPE);
        // 实例化传感器事件监听器
        sensorListener = new SensorListener();
        sensorManager.registerListener(sensorListener, gyroscopeSensor,
                SensorManager.SENSOR_DELAY_UI);
    }
```

```
class SensorListener implements SensorEventListener {
        @Override
        public void onSensorChanged(SensorEvent event) {

                float[] value = event.values;
                // 获取实时的数据并显示
                float x = value[0];
                float y = value[1];
                float z = value[2];
                String info = " x = " + x + "\n y = " + y + "\n z = " + z;
                infoTv.setText(info);

                // 对实时数据与历史最大值信息进行比较，然后进行显示
                if (x > maxX) {
                        maxX = x;
                }
                if (y > maxY) {
                        maxY = y;
                }
                if (z > maxZ) {
                        maxZ = z;
                }
                String stmaxInfo =
                        " x = " + maxX + "\n y = "
                         + maxY + "\n z = "+ maxZ;
                maxInfoTv.setText(stmaxInfo);
        }
        @Override
        public void onAccuracyChanged(Sensor sensor, int accuracy) {

        }
 };
 @Override
 protected void onDestroy() {
        super.onDestroy();
        // 取消监听
        sensorManager.unregisterListener(sensorListener);
 }
}
```

上述代码中，与加速度传感器的代码基本相同，运行程序后，分别沿 X 轴、Y 轴、Z

轴进行快速旋转，观察最大值信息的变化。其中，沿 Y 轴方向做快速旋转操作，效果如图 5-7 所示。

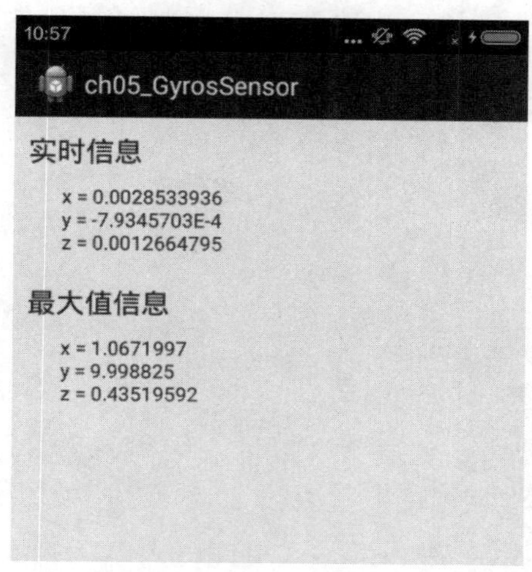

图 5-7　陀螺仪传感器

5.2.7　磁场传感器

磁场传感器(Sensor.TYPE_MAGNETIC_FIELD)用于读取设备周边磁场的变化，可以通过该传感器实现与指南针等磁场有关的程序。磁场传感器同样也具有空间直角坐标系 X 轴、Y 轴、Z 轴三个方向的磁场分量值，单位为 uT(微特斯拉)。

通过 X 轴、Y 轴、Z 轴可以计算磁场数值，公式如下：

$$M = \sqrt{x^2 + y^2 + z^2}$$

下述代码用于实现：通过编写程序实现磁场传感器中 X 轴、Y 轴、Z 轴所对应的分量值的显示，并计算出磁场数值。

(1) 创建新项目(ch05_MagneSensor)，编写布局文件“activity_main.xml”，代码如下：

```
<LinearLayout xmlns:android="http://schemas.android.com/apk/res/android"
    xmlns:tools="http://schemas.android.com/tools"
    android:layout_width="match_parent"
    android:layout_height="match_parent"
    android:background="#ededed"
    android:orientation="vertical"
    android:padding="10dp" >

<TextView
        android:id="@+id/act_main_info_tv"
        android:layout_width="wrap_content"
        android:layout_height="wrap_content"
```

```
        android:layout_marginLeft="20dp"
        android:layout_marginTop="10dp" />

</LinearLayout>
```

(2) 编写“MainActivity.java”类，代码如下：

```
public class MainActivity extends Activity {
        private TextView infoTv = null;

        /** 传感器管理器 */
        private SensorManager sensorManager;
        /** 磁场传感器 */
        private Sensor magneticSensor;
        private SensorEventListener sensorListener;

        @Override
        protected void onCreate(Bundle savedInstanceState) {
                super.onCreate(savedInstanceState);
                setContentView(R.layout.activity_main);

                infoTv = (TextView) findViewById(R.id.act_main_info_tv);

                sensorManager = (SensorManager) getSystemService(SENSOR_SERVICE);
                magneticSensor = sensorManager
                                .getDefaultSensor(Sensor.TYPE_MAGNETIC_FIELD);
                // 实例化传感器事件监听器
                sensorListener = new SensorListener();
                sensorManager.registerListener(sensorListener, magneticSensor,
                                SensorManager.SENSOR_DELAY_UI);
        }

        class SensorListener implements SensorEventListener {

                @Override
                public void onSensorChanged(SensorEvent event) {

                        float[] value = event.values;
                        float x = value[0];
                        float y = value[1];
                        float z = value[2];
                        //计算磁场数值
```

```
                double magneticV = Math.sqrt(x * x + y * y + z * z);

                String info = " x 轴分量：" + x
                              + "\n y 轴分量：" + y
                              + "\n z 轴分量："
                             + z + "\n\n 磁场数值：" + magneticV;
                infoTv.setText(info);
            }

            @Override
            public void onAccuracyChanged(Sensor sensor, int accuracy) {

            }
        };

        @Override
        protected void onDestroy() {
            super.onDestroy();
            // 取消监听
            sensorManager.unregisterListener(sensorListener);
        }
}
```

上述代码中，通过磁场传感器返回的三个轴上的数据，计算出磁场数值，运行效果如图 5-8 所示。

图 5-8　磁场传感器

5.2.8　相对湿度传感器

相对湿度传感器(Sensor.TYPE_RELATIVE_HUMIDITY)是空气中水蒸气的含量相比于指定温度下空气中能含有的最大水蒸气量的值，取值范围通常为 0%～100%。配备这种传感器的手机设备较为少见。本书不做讲解。

5.2.9 环境温度传感器

环境温度传感器(Sensor.TYPE_AMBIENT_TEMPERATURE)用于检测当前环境温度，单位为摄氏度(℃)，手机设备很少配备该传感器。本书不做讲解。

5.2.10 旋转矢量传感器

旋转矢量传感器(Sensor.TYPE_ROTATION_VECTOR)属于合成传感器，用于计算相对于设备坐标系的全局坐标系的旋转角度，同时还使用了加速度传感器、磁场传感器和陀螺仪传感器。

该传感器所获得的数据类似于四元数的形式，也就是说 SensorEvent.values[]中有四个值，是一种可选的旋转表示法。

- ✧ values[0]：沿 x 轴方向的旋转矢量正弦分量(x * sin(θ /2))。
- ✧ values[1]：沿 y 轴方向的旋转矢量正弦分量(y * sin(θ /2))。
- ✧ values[2]：沿 z 轴方向的旋转矢量正弦分量(z * sin(θ /2))。
- ✧ values[3]：旋转矢量余弦分量(cos(θ /2))。

旋转矢量传感器的使用方法非常灵活，可用于与运动相关的检测当中，例如手势、角度变化、相对方位的检测等。

5.2.11 重力传感器

重力传感器(Sensor.TYPE_GRAVITY)也是 Android 中比较常用的一种传感器，主要用于监测手机的放置状态，例如屏幕向上，置于平面，或屏幕向下，置于平面。该传感器有三个数据值，分别表示空间直角坐标系中的 X 轴、Y 轴、Z 轴。类似于加速度传感器，但取值范围值通常为−9.8～9.8m/esc²。

X 轴、Y 轴、Z 轴的取值对手机的放置状态影响如下：

- ✧ 当 x 趋向于 0，且 y 绝对值趋向于 9.8 时，手机垂直于平面，且 X 轴平行于平面；当 y 为正数时，手机听筒方向朝上，反之朝下。
- ✧ 当 x 趋向于 9.8，且 y 绝对值趋向于 0 时，手机垂直于平面，且 Y 轴平行于平面；当 x 为正数时，手机屏幕朝向左侧，反之朝向右侧。
- ✧ 当 z 绝对值趋向于 9.8 时，Z 轴平行于平面(手机平方于平面)；当 z 为正数时，屏幕向上，反之屏幕向下。

下述代码用于实现：通过编写程序实现重力传感器的使用，通过获取的 X 轴、Y 轴、Z 轴上的值，观察对应的手机放置状态，以加深对重力传感器的了解。

(1) 创建新项目(ch05_OrientSensor)，编写布局文件“activity_main.xml”，代码如下：

```
<LinearLayout xmlns:android="http://schemas.android.com/apk/res/android"
    xmlns:tools="http://schemas.android.com/tools"
    android:layout_width="match_parent"
    android:layout_height="match_parent"
```

```
    android:background="#ededed"
    android:orientation="vertical"
    android:padding="10dp" >

<TextView
        android:id="@+id/act_main_info_tv"
        android:layout_width="wrap_content"
        android:layout_height="wrap_content"
        android:layout_marginLeft="20dp"
        android:layout_marginTop="10dp"
        android:textSize="20sp" />

</LinearLayout>
```

(2) 编写“MainActivity.java”类，代码如下：

```
public class MainActivity extends Activity {

        private TextView infoTv = null;

        /** 传感器管理器 */
        private SensorManager sensorManager;
        /** 重力传感器 */
        private Sensor gravitySensor;
        private SensorEventListener sensorListener;

        @Override
        protected void onCreate(Bundle savedInstanceState) {
                super.onCreate(savedInstanceState);
                setContentView(R.layout.activity_main);

                infoTv = (TextView) findViewById(R.id.act_main_info_tv);

                sensorManager = (SensorManager) getSystemService(SENSOR_SERVICE);
                gravitySensor = sensorManager.getDefaultSensor(Sensor.TYPE_GRAVITY);
                // 实例化传感器事件监听器
                sensorListener = new SensorListener();
                sensorManager.registerListener(sensorListener, gravitySensor,
                                SensorManager.SENSOR_DELAY_UI);
        }

        class SensorListener implements SensorEventListener {
```

```
        @Override
        public void onSensorChanged(SensorEvent event) {

            float[] value = event.values;
            float x = value[0];
            float y = value[1];
            float z = value[2];

            String info = " x 轴= " + x
                    + "\n y 轴= " + y
                    + "\n z 轴=" + z;
            infoTv.setText(info);

        }

        @Override
        public void onAccuracyChanged(Sensor sensor, int accuracy) {

        }
    };

    @Override
    protected void onDestroy() {
        super.onDestroy();
        // 取消监听
        sensorManager.unregisterListener(sensorListener);
    }
}
```

上述代码中，对重力传感器做了监听，运行程序后，将手机处于不用的放置状态，观察 X 轴、Y 轴、Z 轴上数值的变化，例如手机平放于桌面时，效果如图 5-9 所示。通过重力传感器可以实现来电翻转静音等特色功能。

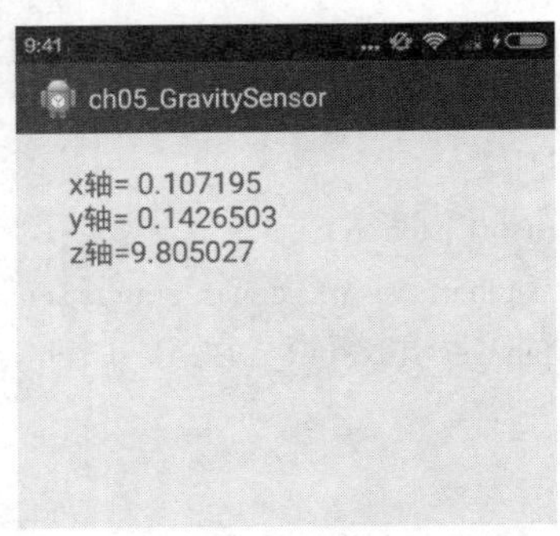

图 5-9　重力传感器

5.2.12　线性加速度传感器

线性加速度传感器(Sensor.TYPE_LINEAR_ACCELERATION)的数据是加速度传感器减去重力影响获取的数据，该传感器与加速度传感器类似，也有三个数据，分别表示空间直角坐标系中 X 轴、Y 轴、Z 轴方向上的分量值，单位为 m/s^2。

线性加速度传感器通常用于获取行进过程中的速度值。该传感器会有一个偏移量，需要把它去掉，方法就是把手机放置在平面上，读取三个轴的数据，然后去掉这些偏移量，就可得到真正的线性加速度值。

5.2.13　方向传感器

方向传感器(Sensor.TYPE_ORIENTATION)可以用来做指南针、地平尺等程序。使用方向传感器时，可以从其传感器事件监听器中 onSensorChanged()回调方法的“event.values”数组中获取三个值，这三个值分别是：

- ✧ values[0]：方位角，绕着 Z 轴的旋转角度。取值范围[0,360]，当手机处于屏幕向上时，0 表示正北方向，90 表示正东方向，180 表示正南方向，270 表示正西方向。可以利用“values[0]”的特性，开发指南针程序。
- ✧ values[1]：倾斜角，绕着 X 轴的旋转角度。取值范围[−180,180]，当手机处于水平放置时，屏幕向上，此值为 0；屏幕向下，此值为 180。可以利用“values[1]”的特性，开发地平尺程序。
- ✧ values[2]：滚动角，绕着 Y 轴的旋转角度。取值范围[−90,90]，当手机处于水平放置且屏幕向上时，此值为 0；当手机以 Y 轴向右旋转时，此值逐渐变小，直到垂直时，此值为−90；继续向右旋转，值逐渐变大，直到屏幕水平向下时，值为 0。

下述代码用于实现：通过编写程序实现方向传感器中“方位角”“倾斜角”“滚动角”三个值的显示，并通过在真机中运行程序来体验三个值的概念及变化。

(1) 创建新项目(ch05_OrientSensor)，编写布局文件“activity_main.xml”，代码如下：

```
<LinearLayout xmlns:android="http://schemas.android.com/apk/res/android"
    android:layout_width="match_parent"
    android:layout_height="match_parent"
    android:background="#ededed"
    android:orientation="vertical"
    android:padding="10dp" >

<TextView
        android:id="@+id/act_main_orientation_tv"
        android:layout_width="wrap_content"
        android:layout_height="wrap_content"
        android:text="方位角:"
```

```
        android:textSize="20sp" />

    <TextView
        android:id="@+id/act_main_incline_tv"
        android:layout_width="wrap_content"
        android:layout_height="wrap_content"
        android:layout_marginTop="10dp"
        android:text="倾斜角:"
        android:textSize="20sp" />

    <TextView
        android:id="@+id/act_main_scroll_tv"
        android:layout_width="wrap_content"
        android:layout_height="wrap_content"
        android:layout_marginTop="10dp"
        android:text="滚动角:"
        android:textSize="20sp" />

</LinearLayout>
```

(2) 编写“MainActivity.java”类，代码如下：

```
public class MainActivity extends Activity {

    /** 方位角 */
    private TextView orientationTv;
    /** 倾斜角 */
    private TextView inclineTv;
    /** 滚动角 */
    private TextView scrollTv;
    /** 传感器管理器 */
    private SensorManager sensorManager;
    /** 方向传感器 */
    private Sensor orientationSensor;

    private SensorEventListener sensorListener;

    @Override
    protected void onCreate(Bundle savedInstanceState) {
        super.onCreate(savedInstanceState);
        setContentView(R.layout.activity_main);
```

```
            orientationTv =
                    (TextView) findViewById(R.id.act_main_orientation_tv);
            inclineTv = (TextView) findViewById(R.id.act_main_incline_tv);
            scrollTv = (TextView) findViewById(R.id.act_main_scroll_tv);

            sensorManager = (SensorManager) getSystemService(SENSOR_SERVICE);
            orientationSensor = sensorManager
                            .getDefaultSensor(Sensor.TYPE_ORIENTATION);
            // 实例化传感器事件监听器
            sensorListener = new SensorListener();
            sensorManager.registerListener(sensorListener,
                    orientationSensor,SensorManager.SENSOR_DELAY_UI);
    }

    class SensorListener implements SensorEventListener {

            @Override
            public void onSensorChanged(SensorEvent event) {
                    orientationTv.setText("方位角： " + event.values[0]);
                    inclineTv.setText("倾斜角： " + event.values[1]);
                    scrollTv.setText("滚动角： " + event.values[2]);
            }
            @Override
            public void onAccuracyChanged(Sensor sensor, int accuracy) {
            }
    };
    @Override
    protected void onDestroy() {
            super.onDestroy();
            // 取消监听
            sensorManager.unregisterListener(sensorListener);
    }
}
```

上述代码中，实现了对方向传感器的监听，运行程序后会实时显示方向传感器的三个参数，运行结果如图 5-10 所示。

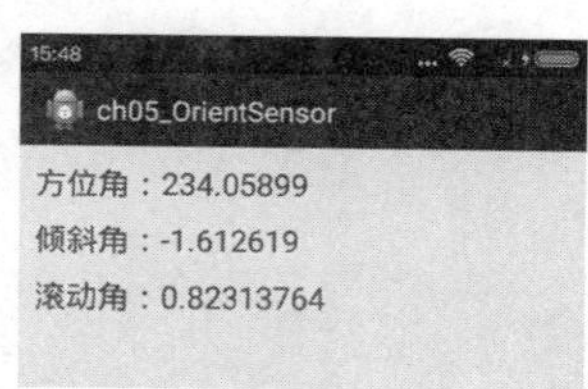

图 5-10　方向传感器

本 章 小 结

(1) Android 中将传感器分为两类：原始传感器和虚拟(复合)传感器。

(2) 原始传感器是指设备中实际存在的物理设备，给出从传感器获得的原始数据。

(3) 虚拟传感器是指，通过多个原始传感器相结合，修改或组合这些原始数据来获得更容易被使用的复合数据。

(4) 通过 SensorManager.getDefaultSensor()方法，传入相应的传感器常量，即可获取手机中相应的传感器。

(5) 使用 SensorEventListener 接口对传感器设置监听，当传感器状态发生变化时，会自动回调此接口中的方法。

本 章 练 习

(1) 使用下列哪个传感器可以实现“拨打电话时关闭屏幕”功能：

(A) 光线传感器　　(B) 温度传感器

(C) 距离传感器　　(D) 加速度传感器

(2) 使用下列哪个传感器可以实现“计算海拔”功能：

(A) 重力传感器　　(B) 气压传感器

(C) 陀螺仪传感器　　(D) 加速度传感器

(3) 通过真机测试，查看当前手机所配备的传感器列表。

(4) 利用重力传感器实现手机翻转静音功能。

第 6 章　Wi-Fi 与 Bluetooth

本章目标

- 理解 Wi-Fi 与 Bluetooth 概念及其实现原理
- 理解管理手机 Wi-Fi 开关以及扫描周围设备
- 理解管理手机 Bluetooth 开关与配对
- 理解通过 Bluetooth 进行设备间通信

本章主要讲解 Wi-Fi 和 Bluetooth(蓝牙)技术。目前这两种技术几乎和智能手机用户密不可分，尤其是 Wi-Fi 无线网络技术，它能够提供高带宽、经济实惠的上网体验；手机中的蓝牙技术通常被应用于蓝牙耳机、智能手环、车载语音等，也可用于设备间文件传输。

6.1 Wi-Fi

Wi-Fi(Wireless Fidelity)又称为“热点”，是一项基于 IEEE 802.11 标准的无线网络连接技术，可以将带有无线网络连接模块的电脑、手持设备等终端，以无线方式互相连接。需要注意的是，Wi-Fi 并不等同于无线网络，它只是无线网络中的一个分支。

6.1.1 Wi-Fi 概述

Android 提供了一套完整的 Wi-Fi 相关 API，使用这些 API 可以对附近的 Wi-Fi 热点进行扫描、连接等操作，同时也可以对本机的 Wi-Fi 模块进行打开、关闭等一系列的配置。其中，主要的类有：

(1) ScanResult：用于描述已检测到的接入点信息，包括接入点的地址、加密方案、频率、信号强度以及网络别名等，ScanResult 类中用以下变量描述这些信息：

- BSSID：接入点的地址，可理解为路由器的 MAC 地址。
- capabilities：描述了身份验证、密钥管理和访问点支持的加密方案。
- frequency：频率，单位为 MHz。
- level：信号强度，单位为 dBm。
- SSID：网络别名。

(2) WifiConfiguration：用于配置无线网络，包括安全配置。

(3) WifiManager：用于管理 Wi-Fi 的连接或关闭，以及对 Wi-Fi 连接的变化进行监听等操作。可以通过 Context.getSystemService()方法获取 WifiManager 对象，具体代码如下：

```
WifiManager manager = (WifiManager)getSystemService(Context.WIFI_SERVICE);
```

该类中定义了用于描述连接状态的一些常量，如表 6-1 所示。

表 6-1 WifiManager 类中用于描述 Wi-Fi 状态的常量

常 量 名	描 述
WIFI_STATE_DISABLED	Wi-Fi 不可用
WIFI_STATE_DISABLING	Wi-Fi 正在关闭
WIFI_STATE_ENABLED	Wi-Fi 可用
WIFI_STATE_ENABLING	Wi-Fi 正在打开
WIFI_STATE_UNKNOWN	未知状态

WifiManager 类中，还提供了用于管理 Wi-Fi 设备操作的方法，常用方法如表 6-2 所示。

表 6-2　WifiManager 类中常用方法

方 法 名	描 述
setWifiEnabled(booleanenabled)	启用或禁用 Wi-Fi 设备
startScan()	扫描 Wi-Fi 设备(热点)
getScanResults()	获取扫描结果，返回值是 List<ScanResult>
getConnectionInfo()	获取 Wi-Fi 连接信息，返回值是 WifiInfo 对象
getConfiguredNetworks()	获取连接过的 Wi-Fi 设备配置列表，返回值是 List<WifiConfiguration>
addNetwork(WifiConfiguration config)	添加一个配置好的 Wi-Fi 设备信息
enableNetwork(int netId, boolean disableOthers)	激活连接指定 Wi-Fi 热点，netId：WifiConfiguration 对象的 id；disableOthers：是否禁用其他已连接的 Wi-Fi 热点
disableNetwork(int netId)	禁用指定 netId 的 Wi-Fi 热点

(4) WifiInfo：用于描述已经建立连接的 Wi-Fi 信息，例如 IP 地址、MAC 地址、信号强度等。该类中常用的方法如表 6-3 所示。

表 6-3　WifiInfo 类中常用方法

方 法 名	描 述
getBSSID()	获取接入点的地址
getHiddenSSID()	SSID 是否被隐藏
getIpAddress()	获取 IP 地址
getLinkSpeed()	获取连接速度
getMacAddress()	获取 Mac 地址
getRssi()	获取 802.11n 网络的信号
getSSID()	获取 SSID(网络别名)
getDetailedStateOf()	获取客户端的连通性

6.1.2　扫描周围的 Wi-Fi

当在 Android 系统的“设置”中打开“WLAN”设置页面后，若 WLAN 为开启状态时，页面下方的列表中会显示出当前扫描到的周围的可用热点。而这里的“WLAN”指的就是“Wi-Fi”功能，“热点”指的是周围的“Wi-Fi”发射器，比如无线路由器、移动热点等设备。同样的，通过代码也可以实现这一功能。

下述示例用于实现：通过编写代码的方式扫描当前环境下的 Wi-Fi 设备，并显示这些设备的部分属性，在扫描之前需要对本机 Wi-Fi 的开启状态进行判断，如果没有开启，则提示用户开启。

(1) 创建新项目(ch06_wifi_list)，编写布局文件“activity_main.xml”，代码如下：

```
<LinearLayout xmlns:android="http://schemas.android.com/apk/res/android"
    xmlns:tools="http://schemas.android.com/tools"
```

```
    android:layout_width="match_parent"
    android:layout_height="match_parent"
    android:background="#ededed"
    android:orientation="vertical"
    android:padding="10dp" >

    <Button
        android:id="@+id/act_main_scan_btn"
        android:layout_width="fill_parent"
        android:layout_height="wrap_content"
        android:text="扫描 Wi-Fi" />

    <ScrollView
        android:layout_width="fill_parent"
        android:layout_height="wrap_content" >

    <TextView
            android:id="@+id/act_main_content_tv"
            android:layout_width="wrap_content"
            android:layout_height="wrap_content"
            android:layout_marginTop="10dp"
            android:textSize="20sp" />
    </ScrollView>

</LinearLayout>
```

(2) 编写“MainActivity.java”类，代码如下：

```
public class MainActivity extends Activity {

    private TextView contentTv = null;
    private Button scanBtn = null;
    private WifiManager wifiManager;

    @Override
    protected void onCreate(Bundle savedInstanceState) {
        super.onCreate(savedInstanceState);
        setContentView(R.layout.activity_main);

        contentTv = (TextView) findViewById(R.id.act_main_content_tv);
        scanBtn = (Button) findViewById(R.id.act_main_scan_btn);
        scanBtn.setOnClickListener(new OnClickListener() {
```

```
            @Override
            public void onClick(View v) {
                scanWifiList();
            }
        });
        wifiManager =
            (WifiManager) getSystemService(Context.WIFI_SERVICE);
}

/**
 * 扫描 Wi-Fi 设备
 *
 * @return
 */
public void scanWifiList() {
    // 如果本机 Wi-Fi 处于关闭状态，则提示用户开启
    if (!wifiManager.isWifiEnabled()) {
        openWiFi();
        return;
    }
    wifiManager.startScan();// 开始扫描
    List<ScanResult> wifiList = wifiManager.getScanResults();

    contentTv.setText("共扫描到 " + wifiList.size()
            + " 个 Wi-Fi 热点：\n\n");

    StringBuffer sb = new StringBuffer();
    for (ScanResult sr : wifiList) {
        sb.append(" SSID: " + sr.SSID)
                .append("\n BSSID: " + sr.BSSID)
                .append("\n capabilities: " + sr.capabilities)
                .append("\n frequency: " + sr.frequency)
                .append("\n level: " + sr.level)
                .append("\n------\n");
    }
    contentTv.append(sb);
}

/**
```

```
        * 打开本机 Wi-Fi
        */
    private void openWiFi() {
            new AlertDialog.Builder(this)
                    .setMessage("Wi-Fi 未开启，是否尝试开启？")
                    .setPositiveButton("开启",
                            new DialogInterface.OnClickListener() {

                                @Override
                                public void onClick(DialogInterface dialog,
                                        int which) {
                                    // 打开本机 Wi-Fi
                                    wifiManager.setWifiEnabled(true);
                                }
                            }).setNegativeButton("取消", null).show();
    }
}
```

上述代码中，主要代码位于 scanWifiList()方法中，用于扫描当前环境下所有 Wi-Fi 热点设备。需要注意的是：无法搜索到匿名的 Wi-Fi 热点；在扫描之前，可以通过 WifiManager.isWifiEnabled()方法获取本机 Wi-Fi 设备的开启状态，如果未开启，可通过调用 WifiManager.setWifiEnabled(true)方法尝试开启。

(3) 在“AndroidManifest.xml”文件中添加 Wi-Fi 相关网络权限，代码如下：

```
<uses-permission android:name="android.permission.CHANGE_WIFI_STATE" />
<uses-permission android:name="android.permission.ACCESS_WIFI_STATE" />
```

(4) 运行程序后，观察所扫描到的 Wi-Fi 热点信息。

在编写 Wi-Fi 相关操作时，必须在“AndroidManifest.xml”文件中声明上述示例中的权限，在之后的小节中，不再过多介绍。

6.1.3 Wi-Fi 相关广播事件

在很多情况下，开发者需要掌控当前 Wi-Fi 设备的开启状态以及 Wi-Fi 网络的连接状态，以进行相关操作。Android 提供了解决方法，当上述事件发生时，系统会自动发送相应的广播，开发者通过注册相应广播，可以“监听”Wi-Fi 网络相关事件。WifiManager 类提供了与 Wi-Fi 相关常用广播的 Action 常量：

- ✧ NETWORK_STATE_CHANGED_ACTION：Wi-Fi 网络连接状态发生变化时，该广播会被多次触发，可以从该广播中获取 NetworkInfo 对象，从而可判断连接状态，这些状态被定义到了 NetworkInfo 类中的 State 枚举中。状态分为：
  - ➢ CONNECTED(已连接)。
  - ➢ CONNECTING(正在连接)。

- DISCONNECTED(已断开)。
- DISCONNECTING(正在断开)。
- SUSPENDED(被暂停)。
- UNKNOWN(未知状态)。

获取 NetworkInfo 对象的代码如下：

```
intent.getParcelableExtra(WifiManager.EXTRA_NETWORK_INFO)
```

✧ RSSI_CHANGED_ACTION：当信号强度发生变化时，可以从该广播中获取当前信号的强度，extra 为 EXTRA_NEW_RSSI，是 Int 类型的值，获取方式如下：

```
Bundle bundle = intent.getExtras();
int rssi = bundle.getInt(WifiManager.EXTRA_NEW_RSSI);
```

✧ SUPPLICANT_CONNECTION_CHANGE_ACTION：当 Wi-Fi 设备被打开或关闭时(与是否连接网络无关)，该广播只触发一次，可以从中获取当前开启状态，extra 为 EXTRA_SUPPLICANT_CONNECTED，是 boolean 类型的值，“true”表示 Wi-Fi 设备已开启，获取方式如下：

```
Bundle bundle = intent.getExtras();
boolean isConnected =
        bundle.getBoolean(WifiManager.EXTRA_SUPPLICANT_CONNECTED);
```

✧ WIFI_STATE_CHANGED_ACTION：当 Wi-Fi 设备状态发生变化时，可以从该广播中获取当前状态和前一次的状态，对应的 extra 分别是：EXTRA_WIFI_STATE 和 EXTRA_PREVIOUS_WIFI_STATE。这些状态使用 WifiManager 中的常量来表示，分别为：

- WIFI_STATE_DISABLED：不可用。
- WIFI_STATE_DISABLING：正在关闭。
- WIFI_STATE_ENABLED：可用。
- WIFI_STATE_ENABLING：正在打开。
- WIFI_STATE_UNKNOWN：未知状态。

获取相应状态的方式如下：

```
Bundle bundle = intent.getExtras();
int crtState = bundle.getInt(WifiManager.EXTRA_WIFI_STATE);
int preState = bundle.getInt(WifiManager.EXTRA_PREVIOUS_WIFI_STATE);
```

接下来将通过具体示例演示以上四种 Wi-Fi 相关广播的使用，这些示例将在同一项目中进行，项目名称为“ch06_wifi_receivers”。

1. NETWORK_STATE_CHANGED_ACTION

下述示例用于实现：通过编写代码的方式来监听 Wi-Fi 网络连接状态的变化情况。

(1) 创建 Activity 类“NetStateChangedActivity.java”，编写代码如下：

```
public class NetStateChangedActivity extends Activity {
    private BroadcastReceiver wifiReceiver = null;
```

```
    @Override
    protected void onCreate(Bundle savedInstanceState) {
        super.onCreate(savedInstanceState);
        TextView tv=new TextView(this);
        tv.setText(
                "正在监听 Wi-Fi 网络连接状态\nNETWORK_STATE_CHANGED_ACTION");
        setContentView(tv);

        initReceiver();
        IntentFilter filter = new IntentFilter(
                    WifiManager.NETWORK_STATE_CHANGED_ACTION);
        registerReceiver(wifiReceiver, filter);
    }

    private void initReceiver() {
        wifiReceiver = new BroadcastReceiver() {

            @Override
            public void onReceive(Context context, Intent intent) {
                NetworkInfo networkInfo = intent
                .getParcelableExtra(WifiManager.EXTRA_NETWORK_INFO);
                String state = "";
                if (null != networkInfo) {
                    switch (networkInfo.getState()) {
                    case CONNECTED:
                        state = "已连接";
                        break;
                    case CONNECTING:
                        state = "正在连接";
                        break;
                    case DISCONNECTED:
                        state = "已断开";
                        break;
                    case DISCONNECTING:
                        state = "正在断开";
                        break;
                    case SUSPENDED:
                        state = "被暂停";
                        break;
                    case UNKNOWN:
```

```
                        state = "未知";
                        break;
                    }
                }
                System.out.println("Wi-fi 网络连接状态：" + state);
            }

        };
    }

    @Override
    protected void onDestroy() {
        unregisterReceiver(wifiReceiver);
        super.onDestroy();
    }
}
```

为简单起见，该 Activity 没有关联 Layout 布局文件，只添加一个 TextView 控件，用于标注当前 Activity 的操作。在 onCreate()方法中注册监听器，在销毁 Activity 时一定要通过 onDestroy()回调方法取消监听。Activity 中的代码是一种对广播接收器的典型使用方式，其他三个广播的使用方式与本示例中的方式类似。

(2) 在“AndroidManifest.xml”文件中注册 Activity，代码如下：

```
<activity android:name=".NetStateChangedActivity" />
```

(3) 运行程序，在 Eclipse 中打开 LogCat 窗口，通过关闭或打开手机的 Wi-Fi，观察输出结果。

2．RSSI_CHANGED_ACTION

下述示例用于实现：通过编写代码的方式来监听 Wi-Fi 网络信号变化的情况。

(1) 创建 Activity 类“RssiChangedActivity.java”，编写代码如下：

```
public class RssiChangedActivity extends Activity {

    private BroadcastReceiver wifiReceiver = null;

    @Override
    protected void onCreate(Bundle savedInstanceState) {
        super.onCreate(savedInstanceState);
        TextView tv = new TextView(this);
        tv.setText("正在监听 Wi-Fi 信号强度\nRSSI_CHANGED_ACTION");
        setContentView(tv);

        initReceiver();
```

```
		IntentFilter filter =
				new IntentFilter(WifiManager.RSSI_CHANGED_ACTION);
		registerReceiver(wifiReceiver, filter);
	}

	private void initReceiver() {
		wifiReceiver = new BroadcastReceiver() {

			@Override
			public void onReceive(Context context, Intent intent) {
				String action = intent.getAction();
				if (action.equals(WifiManager.RSSI_CHANGED_ACTION)) {
					Bundle bundle = intent.getExtras();
					int rssi =
							bundle.getInt(WifiManager.EXTRA_NEW_RSSI);
					System.out.println("Wi-Fi 网络信号强度：" + rssi);
				}
			}

		};
	}

	@Override
	protected void onDestroy() {
		unregisterReceiver(wifiReceiver);
		super.onDestroy();
	}
}
```

(2) 在“AndroidManifest.xml”文件中注册 Activity，代码如下：

```
<activity android:name=".RssiChangedActivity" />
```

(3) 运行程序，在 Eclipse 中打开 LogCat 窗口，连接到 Wi-Fi 网络后，观察输出的信号强度变化。

3. SUPPLICANT_CONNECTION_CHANGE_ACTION

下述示例用于实现：通过编写代码的方式来监听 Wi-Fi 设备开启状态的情况。

(1) 创建 Activity 类“SupplicantConnChangeActivity.java”，编写代码如下：

```
public class SupplicantConnChangeActivity extends Activity {

	private BroadcastReceiver wifiReceiver = null;
```

```
    @Override
    protected void onCreate(Bundle savedInstanceState) {
            super.onCreate(savedInstanceState);
            TextView tv = new TextView(this);
            tv.setText(
            "正在监听 Wi-Fi 设备开启状态\nSUPPLICANT_CONNECTION_CHANGE_ACTION");
            setContentView(tv);

            initReceiver();
            IntentFilter filter = new IntentFilter(
                        WifiManager.SUPPLICANT_CONNECTION_CHANGE_ACTION);
            registerReceiver(wifiReceiver, filter);
    }

    private void initReceiver() {
            wifiReceiver = new BroadcastReceiver() {

                    @Override
                    public void onReceive(Context context, Intent intent) {
                            String action = intent.getAction();
                            if (action.equals(
                            WifiManager.SUPPLICANT_CONNECTION_CHANGE_ACTION)) {
                                    Bundle bundle = intent.getExtras();
                                    boolean isConnected = bundle.getBoolean(
                                            WifiManager.EXTRA_SUPPLICANT_CONNECTED);
                                    System.out.println(
                                            "Wi-fi 设备状态：" + isConnected);
                            }
                    }
            };
    }

    @Override
    protected void onDestroy() {
            unregisterReceiver(wifiReceiver);
            super.onDestroy();
    }
}
```

(2) 在“AndroidManifest.xml”文件中注册该 Activity，代码如下：

```
<activity android:name=".SupplicantConnChangeActivity" />
```

(3) 运行程序，在 Eclipse 中打开 LogCat 窗口，通过关闭或打开手机的 Wi-Fi，观察输出结果。

4. WIFI_STATE_CHANGED_ACTION

下述示例用于实现：通过编写代码的方式来监听 Wi-Fi 设备状态的变化情况。

(1) 创建 Activity 类“WifiStateChangedActivity.java”，编写代码如下：

```
public class WifiStateChangedActivity extends Activity {
	private BroadcastReceiver wifiReceiver = null;

	@Override
	protected void onCreate(Bundle savedInstanceState) {
		super.onCreate(savedInstanceState);
		TextView tv = new TextView(this);
		tv.setText("正在监听 Wi-Fi 设备状态\nWIFI_STATE_CHANGED_ACTION");
		setContentView(tv);

		initReceiver();
		IntentFilter filter = new IntentFilter(
					WifiManager.WIFI_STATE_CHANGED_ACTION);
		registerReceiver(wifiReceiver, filter);
	}

	private void initReceiver() {
		wifiReceiver = new BroadcastReceiver() {

			@Override
			public void onReceive(Context context, Intent intent) {
				String action = intent.getAction();
				if (action.equals(
				WifiManager.WIFI_STATE_CHANGED_ACTION)) {
					Bundle bundle = intent.getExtras();
					int crtState =
						bundle.getInt(WifiManager.EXTRA_WIFI_STATE);
					String state = "";
					switch (crtState) {
					case WifiManager.WIFI_STATE_DISABLED:
						state = "不可用";
						break;
					case WifiManager.WIFI_STATE_DISABLING:
						state = "正在关闭";
						break;
```

```
                        case WifiManager.WIFI_STATE_ENABLED:
                            state = "可用";
                            break;
                        case WifiManager.WIFI_STATE_ENABLING:
                            state = "正在打开";
                            break;
                        case WifiManager.WIFI_STATE_UNKNOWN:
                            state = "未知状态";
                            break;
                        }
                        System.out.println("Wi-fi 状态：" + state);
                    }
                }
            };
        }

        @Override
        protected void onDestroy() {
            unregisterReceiver(wifiReceiver);
            super.onDestroy();
        }
}
```

(2) 在“AndroidManifest.xml”文件中注册 Activity，代码如下：

```
<activity android:name=".WifiStateChangedActivity" />
```

(3) 运行程序，在 Eclipse 中打开 LogCat 窗口，通过关闭或打开手机的 Wi-Fi，观察输出结果。

6.1.4 连接到指定 Wi-Fi 网络

通常可以通过 Android 系统中的“设置”页面来连接或配置指定的 Wi-Fi 网络，但是在某些特定情况下，例如通过 Wi-Fi 网络进行通信，这便要求保证通信的设备连接到同一特定的 Wi-Fi 网络，此时如果能够通过程序内部的设置连接网络，则会大大提高用户体验与程序的可靠性。Android 提供了实现相关功能的 API。

连接指定 Wi-Fi 网络的方式通常分为两种：连接到历史保存过的 Wi-Fi 网络和连接到新的 Wi-Fi 网络。下面将通过示例来介绍具体操作方法，项目名称为“ch06_wifi_conn”。

1．连接到历史保存过的 Wi-Fi 网络

如果需要连接到历史保存过的 Wi-Fi 网络，则首先要获取历史保存的 Wi-Fi 网络列表，然后获取指定网络的“networkId”进行网络连接，连接之前，通常需要把当前连接的网络断开。

下述示例用于实现：通过编写代码的方式获取历史保存的 Wi-Fi 网络列表。如果本机 Wi-Fi 设备处于关闭状态，则提醒用户开启，然后点击列表中的一个 Wi-Fi 网络，尝试连接。

(1) 创建新的 Layout 布局文件“activity_saved_wifi.xml”，编写代码如下：

```
<LinearLayout xmlns:android="http://schemas.android.com/apk/res/android"
    xmlns:tools="http://schemas.android.com/tools"
    android:layout_width="match_parent"
    android:layout_height="match_parent"
    android:background="#ededed"
    android:orientation="vertical"
    android:padding="10dp" >

    <Button
        android:id="@+id/act_main_saved_wifi_btn"
        android:layout_width="fill_parent"
        android:layout_height="wrap_content"
        android:text="查看已保存的 Wi-Fi" />

    <ListView
        android:id="@+id/act_main_list"
        android:layout_width="fill_parent"
        android:layout_height="wrap_content"
        android:layout_marginTop="10dp" />

</LinearLayout>
```

(2) 创建新的 Activity 类“ConnSavedWiFiActivity.java”，编写代码如下：

```
public class ConnSavedWiFiActivity extends Activity {

        private Button getWiFiBtn = null;
        private WifiManager wifiManager;
        private List<WifiConfiguration> wifiConfigList = null;

        private ListView saveWifiLv = null;
        private List<String> saveWifiData = null;

        @Override
        protected void onCreate(Bundle savedInstanceState) {
                super.onCreate(savedInstanceState);
                setContentView(R.layout.activity_saved_wifi);
```

```
        wifiManager =
                (WifiManager) getSystemService(Context.WIFI_SERVICE);

        saveWifiLv = (ListView) findViewById(R.id.act_main_list);
        getWiFiBtn = (Button) findViewById(R.id.act_main_saved_wifi_btn);
        getWiFiBtn.setOnClickListener(new OnClickListener() {

                @Override
                public void onClick(View v) {
                        showSaveWiFiList();
                }
        });

        saveWifiLv.setOnItemClickListener(new OnItemClickListener() {

                @Override
                public void onItemClick(AdapterView<?> parent, View view,
                                int position, long id) {
                        // 首先关闭当前连接的 Wi-Fi
                        WifiInfo crtWifi = wifiManager.getConnectionInfo();
                        if (crtWifi != null) {
                        wifiManager.disableNetwork(crtWifi.getNetworkId());
                        }
                        // 得到当前点击的 Wi-Fi
                        WifiConfiguration cfg = wifiConfigList.get(position);
                        // 连接当前 Wi-Fi
                        wifiManager.enableNetwork(cfg.networkId, true);
                        Toast.makeText(getApplicationContext(),
                        "正在尝试连接 " + cfg.SSID, Toast.LENGTH_SHORT).show();
                }
        });
}

/**
 * 获取历史保存过的 Wi-Fi 列表
 */
private void showSaveWiFiList() {
        // 如果本机 Wi-Fi 设备处于关闭状态，则提示用户开启
        if (!wifiManager.isWifiEnabled()) {
                openWiFi();
```

```
                return;
            }
            saveWifiData = new ArrayList<String>();

            wifiConfigList = wifiManager.getConfiguredNetworks();

            for (WifiConfiguration cfg : wifiConfigList) {
                String networkId = " networkId: " + cfg.networkId;
                String ssid = "\n SSID: " + cfg.SSID;

                String status = "\n status: ";
                if (cfg.status == WifiConfiguration.Status.CURRENT) {
                    status += "当前网络";
                } else {
                    status += "点击以连接";
                }

                saveWifiData.add(networkId + ssid + status);
            }
            ArrayAdapter<String> adapter = new ArrayAdapter<String>(this,
                        android.R.layout.simple_list_item_1, saveWifiData);
            saveWifiLv.setAdapter(adapter);
        }

        /**
         * 打开本机 WiFi 设备
         */
        private void openWiFi() {
            new AlertDialog.Builder(this)
            .setMessage("WiFi 设备未开启，是否尝试开启？")
            .setPositiveButton("开启",
                    new DialogInterface.OnClickListener() {
                    @Override
                    public void onClick(DialogInterface dialog, int which) {
                        // 打开本机 WiFi 设备
                        wifiManager.setWifiEnabled(true);
                    }
            }).setNegativeButton("取消", null).show();
        }
}
```

上述代码，在获取已保存的 Wi-Fi 列表之前，首先判断本机 Wi-Fi 设备的开启状态，之后点击一个 Wi-Fi 网络并进行连接。运行项目可加以验证是否成功连接到指定网络，当然，不要忘记最后在“AndroidManifest.xml”中注册相应的权限以及 Activity。此处不再赘述。

2. 连接到新的 Wi-Fi 网络

相对于连接到历史保存过的 Wi-Fi 网络，连接新的 Wi-Fi 网络稍显复杂，因为这需要提供 SSID(网络别名)和连接密码。通常需要判断 SSID 之前是否连接过，如果连接过，则需把之前的信息删除后再重新连接。简单起见，本小节示例将直接进行连接而不对其进行判断。

(1) 创建新的 Layout 布局文件“activity_new_wifi.xml”，编写代码如下：

```
<LinearLayout xmlns:android="http://schemas.android.com/apk/res/android"
    xmlns:tools="http://schemas.android.com/tools"
    android:layout_width="match_parent"
    android:layout_height="match_parent"
    android:background="#ededed"
    android:orientation="vertical"
    android:padding="10dp" >

    <EditText
        android:id="@+id/act_main_ssid_et"
        android:layout_width="match_parent"
        android:layout_height="wrap_content"
        android:hint="请输入 SSID" />

    <EditText
        android:id="@+id/act_main_pwd_et"
        android:layout_width="match_parent"
        android:layout_height="wrap_content"
        android:hint="请输入密码" />

    <Button
        android:id="@+id/act_main_conn_new_wifi_btn"
        android:layout_width="fill_parent"
        android:layout_height="wrap_content"
        android:text="连接到 Wi-Fi" />

</LinearLayout>
```

(2) 创建新的 Activity 类“ConnNewWiFiActivity.java”，编写代码如下：

```
public class ConnNewWiFiActivity extends Activity {
```

```
    private Button connBtn = null;
    private EditText ssidEt = null;
    private EditText pwdEt = null;
    private WifiManager wifiManager;

    @Override
    protected void onCreate(Bundle savedInstanceState) {
        super.onCreate(savedInstanceState);
        setContentView(R.layout.activity_new_wifi);

        wifiManager=(WifiManager)getSystemService(Context.WIFI_SERVICE);

        connBtn = (Button) findViewById(R.id.act_main_conn_new_wifi_btn);
        ssidEt = (EditText) findViewById(R.id.act_main_ssid_et);
        pwdEt = (EditText) findViewById(R.id.act_main_pwd_et);
        connBtn.setOnClickListener(new OnClickListener() {

            @Override
            public void onClick(View v) {
                if (!wifiManager.isWifiEnabled()) {
                    openWiFi();
                    return;
                }
                String ssid = ssidEt.getText().toString();
                String pwd = pwdEt.getText().toString();
                // 连接 Wi-Fi
                connNewWiFi(ssid, pwd);
            }
        });
    }

    /**
     * 连接到新的 Wi-Fi
     *
     * @param ssid
     * @param pwd
     */
    protected void connNewWiFi(String ssid, String pwd) {
        // 添加新的 Wi-Fi，获取返回的 networkId
        int networkId = addNewWifiConfig(ssid, pwd);
```

```
            if (networkId == -1) {
                    Toast.makeText(this, "创建连接失败",
                            Toast.LENGTH_SHORT).show();
            } else {
                    Toast.makeText(this, "正在尝试连接",
                            Toast.LENGTH_SHORT).show();
                    // 通过 networkId 进行连接
                    wifiManager.enableNetwork(networkId, true);
            }
    }

    /**
     * 添加一个新的 W-Fi 热点
     *
     * @param ssid
     * @param pwd
     * @return 返回 networkId
     */
    public int addNewWifiConfig(String ssid, String pwd) {
            WifiConfiguration wifiCfg = new WifiConfiguration();
            // SSID 和密码需要添加双引号
            wifiCfg.SSID = "\"" + ssid + "\"";
            if (pwd.isEmpty()) {
                    wifiCfg.allowedKeyManagement.set(
                            WifiConfiguration.KeyMgmt.NONE);
            } else {
                    wifiCfg.preSharedKey = "\"" + pwd + "\"";
            }
            wifiCfg.hiddenSSID = false;
            wifiCfg.status = WifiConfiguration.Status.ENABLED;
            return wifiManager.addNetwork(wifiCfg);
    }

    /**
     * 打开本机 Wi-Fi 设备
     */
    private void openWiFi() {
            new AlertDialog.Builder(this)
            .setMessage("WiFi 设备未开启，是否尝试开启？")
            .setPositiveButton("开启",
```

```
                new DialogInterface.OnClickListener() {
                @Override
                public void onClick(DialogInterface dialog, int which) {
                        // 打开本机 Wi-Fi 设备
                        wifiManager.setWifiEnabled(true);
                }
            }).setNegativeButton("取消", null).show();
    }
}
```

在上述代码中，首先通过用户输入的 SSID 和连接密码创建一个新的 WifiConfiguration 对象，其次通过 WifiManager 类的 addNetwork()方法将其添加到网络配置列表中，并返回对应的“networkId”，最后通过“networkId”尝试连接网络。需要注意的是：如果连接的 Wi-Fi 设备不需要密码，则需要在创建 WifiConfiguration 对象时，把密码设置为空。代码如下：

```
wifiCfg.allowedKeyManagement.set(WifiConfiguration.KeyMgmt.NONE);
```

6.1.5 Wi-Fi 技术与设备通信

通过 Wi-Fi 与设备通信技术是指手机端连接到指定 Wi-Fi 热点后，通过 IP 地址及端口号连接设备，建立 Socket 通信连接，进行设备的控制及数据的采集。目前，这项技术被广泛应用于物联网技术领域中。

物联网(Internet of Things，IoT)是一项将物体通过网络进行互相连接，实现物体之间通信以及智能化管理的技术。关于物联网的介绍，请自行查阅相关资料。

移动物联网是智能移动设备与物联网的紧密结合，被广泛运用到日常生活中，例如，远程控制家中电器、远程监控报警、燃气泄漏报警等。

除了通过 Wi-Fi 技术实现控制设备功能，还可以通过 Bluetooth(蓝牙)等技术实现。本小节将通过具体示例：演示手机端如何通过 Wi-Fi 技术与开发板(以下统称为“设备”)进行通信。通过手机端，实现控制设备已有功能模块以及采集数据的功能。设备功能模块及通信协议如表 6-4 所示。

表 6-4 设备功能模块及通信协议

模块名称	通信协议	方向
LED 屏幕	指令：screen 参数：文本 示例：screen:Hello! 描述：手机端发送“Hello!”文本到设备	手机端→设备
LED 灯(4 个)	指令：led1、led2、led3、led4 参数：on(开)、off(关) 示例：led1:on 描述：开启设备 1 号 LED 灯	手机端→设备

续表

模块名称	通信协议	方向
蜂鸣器	指令：buzz 参数：蜂鸣时间(秒) 示例：buzz:2 描述：蜂鸣器蜂鸣 2 秒	手机端→设备
温度、湿度、光照传感器	指令：envir 参数：数值依次为温度、湿度、光照 示例：envir:23;58;88 描述：设备发送环境参数到手机端	设备→手机端
按键(4 个)	指令：key 参数：按键编号 示例：key:1 描述：1 号按键被触发	设备→手机端
振动传感器	指令：shake 参数：无 示例：shake 描述：检测到振动	设备→手机端
磁场传感器	指令：magne 参数：无 示例：magne 描述：检测到强磁	设备→手机端
红外接收器	指令：ir 参数：红外码 示例：ir:A3B245 描述：发送接收到的红外码到手机端	设备→手机端
燃气传感器	指令：gas 参数：无 示例：gas 描述：检测到燃气泄漏	设备→手机端

使用的开发板(设备)如图 6-1 所示，该开发板没有集成 Wi-Fi 和蓝牙模块，使用时，直接将相应模块插入接口即可。

图 6-1　开发板(设备)

将串口 Wi-Fi 模块插到开发板接口上，开发板就可以通过 Wi-Fi 模块创建无线 AP(又称为无线接入点，热点)，手机连接到这个热点后，与之建立 Socket 通信通道即可进行通信。串口 Wi-Fi 模块种类、大小不一，但功能基本相同。本示例用到的串口 Wi-Fi 模块如图 6-2 所示。

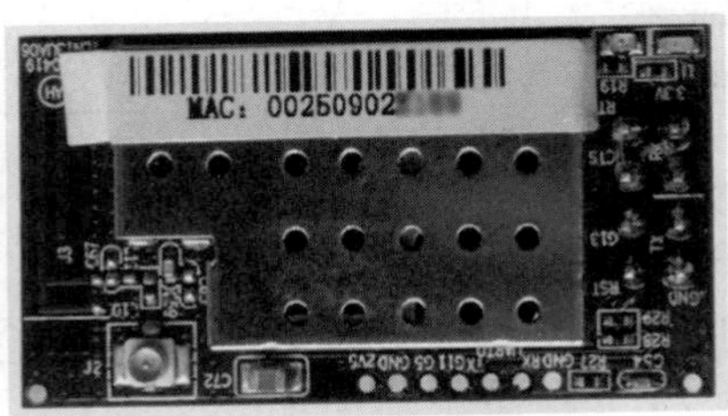

图 6-2　串口 Wi-Fi 模块

注 意

为了简洁易懂，本示例代码中不涉及连接 Wi-Fi 热点的操作，需要通过手机设置程序手动连接 Wi-Fi。

下述示例用于实现：通过用户输入的 IP 地址及端口号与设备创建 Socket 连接，实现对设备的控制，并能够获取设备发送的数据。

(1) 创建项目“ch06_WiFi_IoT”，首先要完成主界面的编写，其主要功能是接收用户输入的 IP 地址及端口号；之后将 IP 地址及端口号传入设备控制界面。修改“activity_main.xml”布局文件，代码如下：

```
<LinearLayout xmlns:android="http://schemas.android.com/apk/res/android"
    xmlns:tools="http://schemas.android.com/tools"
    android:layout_width="match_parent"
    android:layout_height="match_parent"
    android:background="#ededed"
    android:orientation="vertical"
    android:padding="10dp" >

    <EditText
        android:id="@+id/act_main_ip_et"
        android:layout_width="fill_parent"
        android:layout_height="wrap_content"
        android:digits="0123456789."
        android:hint="IP 地址"
        android:singleLine="true"
        android:textSize="20sp" />

    <EditText
        android:id="@+id/act_main_port_et"
        android:layout_width="fill_parent"
```

```
        android:layout_height="wrap_content"
        android:hint="端口号"
        android:inputType="number"
        android:maxLength="5"
        android:singleLine="true"
        android:textSize="20sp" />

    <Button
        android:id="@+id/act_main_control_btn"
        android:layout_width="fill_parent"
        android:layout_height="wrap_content"
        android:layout_marginTop="10dp"
        android:text="打开控制" />

</LinearLayout>
```

(2) 修改“MainActivity.java”，代码如下：

```
public class MainActivity extends Activity {
        private EditText ipEt = null;
        private EditText portEt = null;
        private Button controlBtn = null;
        @Override
        protected void onCreate(Bundle savedInstanceState) {
                super.onCreate(savedInstanceState);
                setContentView(R.layout.activity_main);

                ipEt = (EditText) findViewById(R.id.act_main_ip_et);
                portEt = (EditText) findViewById(R.id.act_main_port_et);
                controlBtn = (Button) findViewById(R.id.act_main_control_btn);
                controlBtn.setOnClickListener(new OnClickListener() {

                        @Override
                        public void onClick(View v) {
                                openControlActivity();
                        }
                });
        }

        private void openControlActivity() {
                if (valiWiFi() && valiInput()) {
                        String ip = ipEt.getText().toString();
```

```
                String port = portEt.getText().toString();
                Intent intent = new Intent(MainActivity.this,
                        ControlActivity.class);
                intent.putExtra("device_ip", ip);
                intent.putExtra("device_port", Integer.parseInt(port));
                startActivity(intent);
            }
        }
        /**
         * 验证网络是否连接到 Wi-Fi 热点
         * @return
         */
        private boolean valiWiFi() {
            ConnectivityManager manager = (ConnectivityManager)
                    getSystemService(Context.CONNECTIVITY_SERVICE);
            State wifiState =
                    manager.getNetworkInfo(ConnectivityManager.TYPE_WIFI)
                            .getState();
            if (wifiState != State.CONNECTED) {
                Toast.makeText(this, "未连接到 Wi-Fi 网络",
                        Toast.LENGTH_SHORT).show();
                return false;
            }
            return true;
        }
        /**
         * 验证输入
         * @return
         */
        private boolean valiInput() {
            String ip = ipEt.getText().toString();
            String port = portEt.getText().toString();
            if (ip.isEmpty() || port.isEmpty()) {
                Toast.makeText(this, "请输入 IP 地址与端口号",
                        Toast.LENGTH_SHORT).show();
                return false;
            }
            return true;
        }
}
```

在上述代码中，当点击“打开控制”按钮时，先后调用 valiWiFi()和 valiInput()方法，分别验证网络是否连接到 Wi-Fi 热点、是否输入了 IP 地址及端口号；之后打开设备控制界面。

(3) 实现设备控制界面，首先创建布局文件“act_control_layout.xml”，代码如下：

```
<RelativeLayout xmlns:android="http://schemas.android.com/apk/res/android"
    xmlns:tools="http://schemas.android.com/tools"
    android:layout_width="match_parent"
    android:layout_height="match_parent"
    android:background="#ededed"
    android:padding="5dp" >

    <LinearLayout
        android:id="@+id/lltop"
        android:layout_width="wrap_content"
        android:layout_height="wrap_content"
        android:orientation="vertical" >

    <TextView
            android:layout_width="fill_parent"
            android:layout_height="wrap_content"
            android:text="环境参数"
            android:textSize="16sp" />

    <TextView
            android:id="@+id/act_ctr_envir_tv"
            android:layout_width="wrap_content"
            android:layout_height="wrap_content"
            android:text="温度：0℃ - 湿度：0% - 光照强度：0"
            android:textSize="16sp" />
    </LinearLayout>

    <LinearLayout
        android:id="@+id/llcenter"
        android:layout_width="fill_parent"
        android:layout_height="fill_parent"
        android:layout_above="@+id/llbottom"
        android:layout_below="@+id/lltop"
        android:orientation="vertical" >

    <TextView
```

```
        android:layout_width="fill_parent"
        android:layout_height="wrap_content"
        android:text="事件触发"
        android:textSize="16sp" />

<ScrollView
        android:id="@+id/act_ctr_scroll"
        android:layout_width="fill_parent"
        android:layout_height="fill_parent" >

<TextView
            android:id="@+id/act_ctr_event_tv"
            android:layout_width="fill_parent"
            android:layout_height="fill_parent"
            android:textSize="16sp" />
</ScrollView>
</LinearLayout>

<LinearLayout
    android:id="@+id/llbottom"
    android:layout_width="fill_parent"
    android:layout_height="wrap_content"
    android:layout_alignParentBottom="true"
    android:orientation="vertical" >

<Button
        android:id="@+id/act_ctr_screen_btn"
        android:layout_width="fill_parent"
        android:layout_height="wrap_content"
        android:text="发送屏显" />

<Button
        android:id="@+id/act_ctr_buzz_btn"
        android:layout_width="fill_parent"
        android:layout_height="wrap_content"
        android:text="蜂鸣器报警" />

<TextView
        android:layout_width="fill_parent"
        android:layout_height="wrap_content"
```

```
            android:layout_marginTop="5dp"
            android:text="LED 灯控制"
            android:textSize="16sp" />

    <LinearLayout
            android:layout_width="wrap_content"
            android:layout_height="wrap_content"
            android:layout_gravity="center_horizontal"
            android:orientation="horizontal" >

        <Switch
                android:id="@+id/act_ctr_led1_sw"
                android:layout_width="wrap_content"
                android:layout_height="wrap_content"
                android:layout_marginTop="15dp"
                android:layout_weight="1"
                android:switchMinWidth="2dp"
                android:switchPadding="2dp"
                android:text="1 号灯"
                android:textOff="关"
                android:textOn="开" />

        <Switch
                android:id="@+id/act_ctr_led2_sw"
                android:layout_width="wrap_content"
                android:layout_height="wrap_content"
                android:layout_marginLeft="20dp"
                android:layout_marginTop="15dp"
                android:layout_weight="1"
                android:switchMinWidth="0dp"
                android:switchPadding="2dp"
                android:text="2 号灯"
                android:textOff="关"
                android:textOn="开" />
    </LinearLayout>

    <LinearLayout
            android:layout_width="wrap_content"
            android:layout_height="wrap_content"
            android:layout_gravity="center_horizontal"
```

```
        android:orientation="horizontal" >

    <Switch
            android:id="@+id/act_ctr_led3_sw"
            android:layout_width="wrap_content"
            android:layout_height="wrap_content"
            android:layout_marginTop="15dp"
            android:layout_weight="1"
            android:switchMinWidth="2dp"
            android:switchPadding="2dp"
            android:text="3 号灯"
            android:textOff="关"
            android:textOn="开" />

    <Switch
            android:id="@+id/act_ctr_led4_sw"
            android:layout_width="wrap_content"
            android:layout_height="wrap_content"
            android:layout_marginLeft="20dp"
            android:layout_marginTop="15dp"
            android:layout_weight="1"
            android:switchMinWidth="2dp"
            android:switchPadding="2dp"
            android:text="4 号灯"
            android:textOff="关"
            android:textOn="开" />
</LinearLayout>
</LinearLayout>
</RelativeLayout>
```

该界面主要用于实现显示设备检测到的环境信息、事件触发记录，以及控制设备等功能。

(4) 创建“ControlActivity.java”类，该类为程序核心类，代码将分步骤实现。首先编写基本代码，包括布局文件中控件的引用等，代码如下：

```
public class ControlActivity extends Activity {
    /** 系统消息 */
    private final static int TAG_WHAT_SYS = 1;
    /** 设备发送来的消息 */
    private final static int TAG_WHAT_DEVICE = 2;

    /** Socket 通信对象 */
```

```
private Socket socket;
/** 手机端通信线程 */
private SocketThread socketThread;

private ScrollView scrollView = null;
/** 环境信息 */
private TextView envirInfoTv = null;
/** 事件信息 */
private TextView eventInfoTv = null;
/** 发送屏显 */
private Button screenBtn = null;
/** 蜂鸣器报警 */
private Button buzzBtn = null;
/** 控制四个 LED 灯 */
private Switch led1Sw = null;
private Switch led2Sw = null;
private Switch led3Sw = null;
private Switch led4Sw = null;

@Override
protected void onCreate(Bundle savedInstanceState) {
        super.onCreate(savedInstanceState);
        setContentView(R.layout.act_control_layout);

        scrollView = (ScrollView) findViewById(R.id.act_ctr_scroll);
        envirInfoTv = (TextView) findViewById(R.id.act_ctr_envir_tv);
        eventInfoTv = (TextView) findViewById(R.id.act_ctr_event_tv);
        screenBtn = (Button) findViewById(R.id.act_ctr_screen_btn);
        buzzBtn = (Button) findViewById(R.id.act_ctr_buzz_btn);
        led1Sw = (Switch) findViewById(R.id.act_ctr_led1_sw);
        led2Sw = (Switch) findViewById(R.id.act_ctr_led2_sw);
        led3Sw = (Switch) findViewById(R.id.act_ctr_led3_sw);
        led4Sw = (Switch) findViewById(R.id.act_ctr_led4_sw);
        led1Sw.setTag("led1");
        led2Sw.setTag("led2");
        led3Sw.setTag("led3");
        led4Sw.setTag("led4");

        screenBtn.setOnClickListener(onBtnClickListener);
        buzzBtn.setOnClickListener(onBtnClickListener);
```

```
            led1Sw.setOnCheckedChangeListener(onSwitchChangeListener);
            led2Sw.setOnCheckedChangeListener(onSwitchChangeListener);
            led3Sw.setOnCheckedChangeListener(onSwitchChangeListener);
            led4Sw.setOnCheckedChangeListener(onSwitchChangeListener);
            // 初始化连接
            initConn();
        }
}
```

上述代码，主要实现了对布局控件的引用。其中，4 个 Switch 开关控件用于控制设备 4 个 LED 灯的开关，将指令名称以 Tag 属性的形式分别添加到这 4 个控件中；“eventInfoTv”对象是事件信息文本框，用于显示事件的相关信息，被放置于 ScrollView 控件中，支持上下滑动。

接下来实现与设备建立 Socket 通信的功能，代码如下：

```
/**
 * 初始化连接
 */
private void initClient() {
        String ip = getIntent().getStringExtra("device_ip");
        int port = getIntent().getIntExtra("device_port", 0);
        String str = "系统：正在尝试连接到设备";
        eventInfoTv.append(str + "\n");
        // 创建客户端线程
        socketThread = new SocketThread(ip, port);
        socketThread.start();
}


/**
 * 手机端通信端线程
 */
private class SocketThread extends Thread {
        String ip = null;
        int port = 0;
        public SocketThread(String ip, int port) {
                super();
                this.ip = ip;
                this.port = port;
        }
        public void run() {
                InputStream in = null;
                socket = new Socket();
```

```
            try {
                    SocketAddress address = new InetSocketAddress(ip, port);
                    // SocketAddress address = new
                    // InetSocketAddress("192.168.1.123",
                    // 5678);
                    socket.connect(address, 3000);
                    sendMsgToHandler(TAG_WHAT_SYS, "系统：连接成功！");

                    // 等待读取服务器发送的数据
                    byte[] buffer = new byte[128];
                    in = socket.getInputStream();
                    int c = 0;
                    while ((c = in.read(buffer)) > 0) {
                            String msg = new String(buffer, 0, c, "gbk");
                            sendMsgToHandler(TAG_WHAT_DEVICE, msg);
                    }
            } catch (IOException e) {
                    sendMsgToHandler(TAG_WHAT_SYS, "系统：连接失败！");
                    e.printStackTrace();
            } finally {
                    try {
                            if (in != null) {
                                    in.close();
                            }
                    } catch (IOException e) {
                            e.printStackTrace();
                    }
            }
    }
};
```

在上述代码中，创建了一个内部类 ClientThread，用于开启子线程，与设备建立 Socket 连接，并对设备发来的数据进行监听处理。当监听到设备发来的数据时，通过 sendMsgToHandler(TAG_WHAT_DEVICE, msg)方法将数据发送到 Handler 进行处理。处理设备发来的数据流程的代码如下：

```
/**
 * 发送消息到 Handler
 *
 * @param what
 * @param msgStr
 */
```

```
public void sendMsgToHandler(int what, final String msgStr) {
        Message msg = new Message();
        msg.what = what;
        msg.obj = msgStr;
        handler.sendMessage(msg);
}

Handler handler = new Handler() {
        @Override
        public void handleMessage(Message msg) {
                String str = (String) msg.obj;
                switch (msg.what) {
                case TAG_WHAT_DEVICE:
                        handleMsgForDevice(str);
                        break;
                case TAG_WHAT_SYS:
                        eventInfoTv.append(str + "\n");
                        break;
                }
        }
};

/**
 * 处理设备发送的消息
 *
 * @param msg
 */
private void handleMsgForDevice(String msgStr) {
        String[] cmdStr = msgStr.split(":");
        String event = cmdStr[0];
        if (event.equals("envir")) {
                String[] envirs = cmdStr[1].split(";");
                String temp = envirs[0];
                String humi = envirs[1];
                String light = envirs[2];
                String msg = "温度：" + temp + "℃ - 湿度：" + humi + "% - 光照强度：" + light;
                envirInfoTv.setText(msg);
        } else if (event.equals("key")) {
                String num = cmdStr[1];
                eventInfoTv.append(num + " 号按键被触发\n");
```

```
    } else if (event.equals("shake")) {
            eventInfoTv.append("检测到振动\n");
    } else if (event.equals("magne")) {
            eventInfoTv.append("检测到强磁\n");
    } else if (event.equals("ir")) {
            String ir = cmdStr[1];
            eventInfoTv.append("接收到红外码：" + ir + "\n");
    } else if (event.equals("gas")) {
            eventInfoTv.append("检测到燃气泄漏\n");
    }
    scrollToLast();
}
```

在上述代码中，sendMsgToHandler()方法负责将传入的消息封装成 Message 对象，发送到 Handler 进行处理；handleMsgForDevice()方法负责解析传入的指令协议，将解析后的内容更新到事件信息文本框中，通过调用 scrollToLast()方法，使事件信息文本框始终显示最后一条信息。scrollToLast()方法代码如下：

```
private void scrollToLast() {
    // ScrollView 自动滚动到最后一行
    scrollView.post(new Runnable() {
        public void run() {
            scrollView.scrollTo(0, 5000);
        }
    });
}
```

至此，单向监听设备发送信息的功能已实现，接下来实现向设备发送指令、控制设备模块的功能。

创建 Button 按钮和 Switch 开关控件的事件监听，代码如下：

```
/** 按钮点击事件 */
private OnClickListener onBtnClickListener = new OnClickListener() {
    @Override
    public void onClick(View v) {
        if (v == screenBtn) {
            showSendScreenMsgDialog();//显示发送文本对话框
        } else if (v == buzzBtn) {
            sendBuzzer();//发送蜂鸣器报警指令
        }
    }
};
/** Switch 开关改变事件 */
private OnCheckedChangeListener onSwitchChangeListener =
```

```
    new OnCheckedChangeListener() {
        @Override
        public void onCheckedChanged(CompoundButton v, boolean checked) {
                String tag = (String) v.getTag();
                buildCmdDataToDevice(tag, checked ? "on" : "off");
        }
};
```

实现 showSendScreenMsgDialog()和 sendBuzzer()方法，代码如下：

```
/**
 * 显示发送屏显消息对话框
 */
protected void showSendScreenMsgDialog() {
        LinearLayout llayout = new LinearLayout(this);
        LinearLayout.LayoutParams ll_lpara = new LinearLayout.LayoutParams(
                    LayoutParams.FILL_PARENT, LayoutParams.WRAP_CONTENT);
        final EditText msgEt = new EditText(this);
        msgEt.setHint("请输入显示文本");
        llayout.addView(msgEt, ll_lpara);
        new AlertDialog.Builder(this).setTitle("发送屏显内容").setView(llayout)
            .setPositiveButton("发送", new DialogInterface.OnClickListener() {
                @Override
                public void onClick(DialogInterface dialog, int id) {
                        String msg = msgEt.getText().toString();
                        buildCmdDataToDevice("screen", msg);
                }
        }).setNeutralButton("取消", null).show();
}

/**
 * 发送蜂鸣器报警
 */
protected void sendBuzzer() {
        buildCmdDataToDevice("buzz", "2");
}
```

接下来，实现 buildCmdDataToDevice()方法，该方法负责将传入的指令和参数来构建最终的指令协议，并把协议发送到设备上，代码如下：

```
/**
 * 构建指令并发送指令到设备
 *
 * @param event
```

```
 * @param param
 */
private void buildCmdDataToDevice(String event, String param) {
        String cmd = event + ":" + param;
        try {
                OutputStream os = socket.getOutputStream();
                os.write(cmd.getBytes("gbk"));
                eventInfoTv.append("发送 " + event + " 指令成功\n");
        } catch (IOException e) {
                e.printStackTrace();
                eventInfoTv.append("发送 " + event + " 指令失败\n");
        }
        scrollToLast();
}
```

在上述代码中，通过 Socket.getOutputStream()方法获取输出流对象，然后将构建好的指令协议发送到设备上。

最后，在当前界面关闭前断开连接，并关闭线程，代码如下：

```
@Override
protected void onDestroy() {
        // 断开连接 ，关闭线程
        if (socket != null) {
                try {
                        socket.close();
                } catch (IOException e) {
                        e.printStackTrace();
                }
                socket = null;
        }
        if (socketThread != null) {
                socketThread.interrupt();
                socketThread = null;
        }
        super.onStop();
}
```

至此，设备控制界面已完成。

(5) 在“AndroidManifest.xml”配置文件中注册“ControlActivity.java”类，代码省略。

(6) 在“AndroidManifest.xml”配置文件中声明必要的权限，代码如下：

```
<uses-permission android:name="android.permission.INTERNET" />
<uses-permission android:name="android.permission.ACCESS_WIFI_STATE" />
<uses-permission android:name="android.permission.ACCESS_NETWORK_STATE" />
```

(7) 运行程序后，连接设备进行测试，手机端效果如图 6-3 所示。

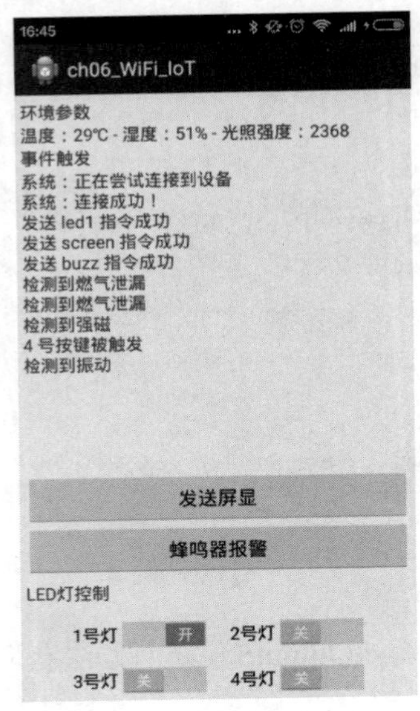

图 6-3　手机端通信界面

6.2　Bluetooth(蓝牙)

从 Android4.3 版本开始，支持 Bluetooth4.0(蓝牙)通信技术。该技术是一种无线技术标准，工作在 2.4 GHz 频段，能够实现设备间短距离、低带宽、低功耗、点对点的局域网通信功能。相对于蓝牙 3.0 版本，新的 4.0 版本具有更低功耗、更小延迟、更长连接距离等优点。随着移动物联网技术浪潮的到来，蓝牙通信技术已被广泛应用于物联网设备中。

蓝牙 4.0 有两个分支：传统蓝牙 4.0 和 BLE4.0。传统蓝牙 4.0 是从之前版本升级而来，支持向下兼容。而 BLE4.0 是一项新的技术，不支持向下兼容。本节将介绍蓝牙通信的相关技术——传统蓝牙通信技术和 BLE 蓝牙通信技术。通过手机之间蓝牙通信示例讲解传统蓝牙通信技术；通过手机与开发板之间蓝牙通信示例讲解 BLE 蓝牙通信技术。

6.2.1　传统蓝牙概述

Android API 的“android.bluetooth”包中对蓝牙技术的实现提供了支持，接下来将对常用的类进行详细介绍。

1．BluetoothAdapter

BluetoothAdapter(蓝牙适配器)，通过该类可以对当前设备的蓝牙进行基本的操作，包括：打开或关闭蓝牙、启用设备发现以及检测蓝牙设备状态等。获取 BluetoothAdapter 对象，代码如下：

```
BluetoothAdapter adapter=BluetoothAdapter.getDefaultAdapter();
```

BluetoothAdapter 类中封装了用于描述本地蓝牙设备当前状态的相关常量，如表 6-5 所示。

表 6-5　BluetoothAdapter 类中状态相关常量

常 量 名	描 述
STATE_OFF	蓝牙设备处于关闭状态
STATE_TURNING_ON	蓝牙设备处于正在打开状态
STATE_ON	蓝牙设备处于开启状态
STATE_TURNING_OFF	蓝牙设备处于正在关闭状态
SCAN_MODE_NONE	无功能状态，蓝牙设备既不能扫描其他设备，也处于不可见状态
SCAN_MODE_CONNECTABLE	蓝牙设备处于扫描状态，可以扫描其他设备，仅对已配对设备可见
SCAN_MODE_CONNECTABLE_DISCOVERABLE	蓝牙设备处于可见状态，既可以扫描其他设备，也可被其他设备发现

当本地蓝牙设备相关属性或状态发生变化时，系统将发送相应的广播通知所有已注册该广播的对象。这些广播的 Action 被定义到 BluetoothAdapter 类中，本地蓝牙相关广播的 Action 常量如表 6-6 所示。

表 6-6　本地蓝牙相关广播的 Action 常量

常 量 名	描 述
ACTION_LOCAL_NAME_CHANGED	当前设备的蓝牙适配器别名被改变。附加值为 EXTRA_LOCAL_NAME，表示当前名称
ACTION_SCAN_MODE_CHANGED	蓝牙扫描模式发生变化。包含了两个附加值，分别为 EXTRA_SCAN_MODE(当前扫描模式)、EXTRA_PREVIOUS_SCAN_MODE(之前的扫描模式)
ACTION_STATE_CHANGED	蓝牙设备开关状态改变。包含了两个附加值，分别为 EXTRA_STATE(当前开关状态)、EXTRA_PREVIOUS_STATE(之前的开关状态)
ACTION_DISCOVERY_STARTED	开始扫描周围蓝牙设备
ACTION_DISCOVERY_FINISHED	完成扫描周围设备操作

BluetoothAdapter 类中关于蓝牙相关操作的常用方法如表 6-7 所示。

表 6-7　BluetoothAdapter 类中的常用方法

方 法 名	描 述
getDefaultAdapter()	是一个静态方法，获取当前设备默认的蓝牙适配器对象
getState()	获取蓝牙设备的状态
enable()	强制打开蓝牙设备，返回 boolean 类型
disable()	关闭蓝牙设备，返回 boolean 类型
isEnable()	蓝牙设备是否可用，返回 boolean 类型

续表

方 法 名	描 述
startDiscovery()	开始扫描周围设备，返回 boolean 类型
cancelDiscovery()	取消扫描周围设备，返回 boolean 类型
isDiscovering()	是否正在扫描，返回 boolean 类型
getScanMode()	获取当前扫描模式
getName()	获取当前设备蓝牙名称
getAddress()	获取当前设备蓝牙 MAC 地址
checkBluetoothAddress(String address)	检查蓝牙 MAC 地址是否合法
getBoundedDevices()	获取当前已配对的蓝牙设备集合，返回 Set<BluetoothDevice>集合对象
getRemoteDevice(String address)	根据 MAC 地址获取周围的蓝牙设备，返回 BluetoothDevice 对象
listenUsingRfcommWithServiceRecord(String name, UUID uuid)	通过服务器名称及 UUID 创建 Rfcommon 端口的蓝牙监听，返回 BluetoothServerSocket 对象

2. BluetoothDevice

BluetoothDevice 类用于描述远端的蓝牙设备，可以通过该类的对象创建一个连接，以及查询远端蓝牙设备的名称、地址和连接状态等。该类的属性无法被修改。可以通过 BluetoothAdapter 对象的 getRemoteDevice(String address)方法获取 BluetoothDevice 对象，参数“address”表示远端蓝牙设备的 MAC 地址；或者从获取的周围蓝牙设备列表(Set<BluetoothDevice>集合对象)中得到。

BluetoothDevice 类中封装了用于描述远端蓝牙设备当前状态的相关常量，如表 6-8 所示。

表 6-8 BluetoothDevice 类中状态常量

常 量 名	描 述
BOND_BONDED	已经与远端蓝牙设备配对
BOND_BONDING	正在与远端蓝牙设备配对
BOND_NONE	远端蓝牙设备未配对

与本地蓝牙设备相同，当远端蓝牙设备相关属性或状态发生变化时，系统也会发送相应的广播。这些广播的 Action 被定义到 BluetoothDevice 类中。远端蓝牙相关广播的 Action 常量如表 6-9 所示。

表 6-9 远端蓝牙相关广播的 Action 常量

常 量 名	描 述
ACTION_ACL_CONNECTED	与远端设备建立低级别(ACL)连接成功。包含一个附加值：EXTRA_DEVICE(BluetoothDevice 远端蓝牙设备)
ACTION_ACL_DISCONNECTED	与远端设备建立低级别(ACL)连接断开。包含一个附加值：EXTRA_DEVICE

续表

常　量　名	描　　述
ACTION_ACL_DISCONNECT_REQUESTED	远端设备请求断开低级别(ACL)连接，并且即将断开。包含一个附加值：EXTRA_DEVICE
ACTION_BOND_STATE_CHANGED	远端设备连接状态已改变。包含两个附加值，分别是 EXTRA_DEVICE、EXTRA_BOND_STATE(当前状态) EXTRA_PREVIOUS_BOND_STATE(之前的状态)
ACTION_CLASS_CHANGED	一个已经改变的远端蓝牙设备类。包含两个附加值，分别是 EXTRA_DEVICE 和 EXTRA_BOND_STATE
ACTION_FOUND	发现远程设备。包含两个附加值，分别是 EXTRA_DEVICE 和 EXTRA_CLASS。如果该设备可用，可获取另外两个附加值，分别是 EXTRA_NAME 或 EXTRA_RSSI
ACTION_NAME_CHANGED	远端设备名称发生变化，或第一次获取到名称。包含两个附加值，分别是 EXTRA_DEVICE 和 EXTRA_NAME

BluetoothDevice 类中关于远端蓝牙相关操作的常用方法如表 6-10 所示。

表 6-10　BluetoothDevice 类中的常用方法

方　法　名	描　　述
createRfcommSocketToServiceRecord(UUID uuid)	通过 UUID 创建一个 Rfcommon 端口，准备一个对远端设备安全的连接，返回值为 BluetoothSocket 对象
describeContents()	获取 Parcelable 中包含的特殊内容对象
getAddress()	获取该蓝牙设备的 MAC 地址
getBluetoothClass()	获取远端蓝牙设备的 BluetoothClass 对象
getBondState()	获取远端蓝牙设备连接状态
getName()	获取远端蓝牙设备的名称
connectGatt(Context context, boolean autoConnect, BluetoothGattCallback callback)	连接到 GATT 协议设备

3. BluetoothSocket

BluetoothSocket 与 Java 的 Socket 类似，用于实现客户端之间的 Socket 通信。该类常用方法如表 6-11 所示。

表 6-11　BluetoothSocket 类中的常用方法

方 法 名	描 述
connect()	与设备建立通信连接
close()	与设备断开连接
getInptuStream()	获取输入流，即读取远端设备数据的流
getOutputStream()	获取输出流，即向远端设备发送数据的流
getRemoteDevice()	获取远端蓝牙设备对象(BluetoothDevice)

4．BluetoothServerSocket

BluetoothServerSocket 是服务器端 Socket 类，该类主要有三个方法，分别是 accept()、accept(int timeout)以及 close()方法。前两个方法为堵塞方法，作用是等待客户端的连接，一旦有客户端连接成功，该方法会返回 BluetoothSocket 对象，之后服务器与客户端的通信都是通过 BluetoothSocket 对象来实现。

6.2.2　传统蓝牙通信

“传统蓝牙”之间通过 Socket 协议进行交换数据，本小节将介绍如何利用蓝牙通信技术实现手机设备间数据的传输。

1．权限

操作蓝牙设备，必须在“AndroidManifest.xml”文件中注册相应的权限。和蓝牙相关的权限有两个：

✧ android.permission.BLUETOOTH：允许程序连接到已配对的蓝牙设备，用于连接或传输数据等配对后的操作。

✧ android.permission.BLUETOOTH_ADMIN：允许程序发现或配对蓝牙设备，针对的是配对前的操作。在注册“android.permission.BLUETOOTH”权限后，该权限才有效。

2．开启蓝牙

在程序中开启蓝牙有两种方式：强制开启和请求开启。

(1) 强制开启，代码如下：

```
BluetoothAdapterbluetoothAdapter = BluetoothAdapter.getDefaultAdapter();
bluetoothAdapter.enable();
```

强制开启，就是不经过用户同意，自动在后台开启蓝牙设备。

(2) 请求开启，代码如下：

```
Intent intent = new Intent(BluetoothAdapter.ACTION_REQUEST_ENABLE);
startActivityForResult(intent, 1);
```

当用请求方式开启蓝牙设备时，系统会以对话框的形式提示用户是否开启蓝牙设备。重写 onActivityResult()方法可获取开启结果，代码如下：

```
@Override
protected void onActivityResult(int requestCode,
```

```
        int resultCode, Intent data) {
            super.onActivityResult(requestCode, resultCode, data);
            if (requestCode == 1) {
                if (resultCode == RESULT_OK) {
                    Toast.makeText(this, "蓝牙开启成功",
                    Toast.LENGTH_SHORT).show();
                } else {
                    Toast.makeText(this, "蓝牙开启失败",
                    Toast.LENGTH_SHORT).show();
                }
            }
}
```

3．使蓝牙设备可见

当本地蓝牙设备希望对远端蓝牙设备可见(能够扫描到本地设备)时，需要向用户发出可见性的请求，代码如下：

```
Intent intent = new Intent(BluetoothAdapter.ACTION_REQUEST_DISCOVERABLE);
intent.putExtra(BluetoothAdapter.EXTRA_DISCOVERABLE_DURATION, 300);
startActivityForResult(intent, 0);
```

在请求中，可以通过添加附加值“EXTRA_DISCOVERABLE_DURATION”修改可见时间，单位为“秒”，默认值为 120 秒；也可以通过重写 onActivityResult()方法获取请求结果。

下述示例用于实现：使用蓝牙通信技术完成设备之间的简单通信。要求：程序同时支持服务器和客户端(远端设备)的功能。程序首页可修改本地蓝牙设备名称、显示历史已配对的设备列表、扫描周围蓝牙设备。当用户点击远端设备列表时，以客户端的形式请求配对并连接远端设备；当用户以服务器方式启动时，开启“可发现模式”，时间为 300 秒；当客户端与服务器成功建立连接后，可进行简单的文本通信。

(1) 创建项目“ch06_bluetooth_chat”，首先编写服务器端的实现。为了提高程序的可读性，把客户端与服务器端通信页面分解为两个相同的页面，使用同一个布局资源。创建“act_bluetooth_chat.xml”布局文件，代码如下：

```
<RelativeLayout xmlns:android="http://schemas.android.com/apk/res/android"
    xmlns:tools="http://schemas.android.com/tools"
    android:layout_width="match_parent"
    android:layout_height="match_parent"
    android:background="#ededed"
    android:orientation="vertical"
    android:padding="10dp" >

<ScrollView
        android:layout_width="fill_parent"
```

```
        android:layout_height="fill_parent"
        android:layout_above="@+id/rl" >

    <TextView
            android:id="@+id/act_chat_content_tv"
            android:layout_width="fill_parent"
            android:layout_height="wrap_content"
            android:text="正在会话：\n"
            android:textSize="16sp" />
</ScrollView>

<RelativeLayout
        android:id="@+id/rl"
        android:layout_width="fill_parent"
        android:layout_height="wrap_content"
        android:layout_alignParentBottom="true" >

    <Button
            android:id="@+id/act_chat_close_btn"
            android:layout_width="wrap_content"
            android:layout_height="wrap_content"
            android:layout_centerVertical="true"
            android:text="关闭" />

<EditText
            android:id="@+id/act_chat_msg_et"
            android:layout_width="fill_parent"
            android:layout_height="wrap_content"
            android:layout_centerVertical="true"
            android:layout_toLeftOf="@+id/act_chat_send_btn"
            android:layout_toRightOf="@+id/act_chat_close_btn"
            android:hint="输入内容" />

    <Button
            android:id="@+id/act_chat_send_btn"
            android:layout_width="wrap_content"
            android:layout_height="wrap_content"
            android:layout_alignParentRight="true"
            android:layout_centerVertical="true"
            android:text="发送" />
```

```
</RelativeLayout>
</RelativeLayout>
```

此布局比较简单，主要分为内容显示区和控制区，控制区可以关闭会话、发送消息。

(2) 实现服务器端的代码编写，主要功能有：

✧ 创建蓝牙通信通道的监听。
✧ 等待客户端的连接。
✧ 读取客户端发送的消息。
✧ 发送消息到客户端。

创建“BluetoothServiceActivity.java”类，首先编写基本代码，包括布局文件中控件的引用等，代码如下：

```
public class BluetoothServiceActivity extends Activity {
    /** 远端蓝牙设备名称 */
    private String remoteName = null;

    private BluetoothAdapter bluetoothAdapter = null;
    /** 服务器 Socket 对象 */
    private BluetoothServerSocket serverSocket = null;
    /** 与客户端通信的 Socket 对象*/
    private BluetoothSocket socket = null;
    /** 服务器线程 */
    private ServerThread serverThread = null;

    private TextView contentTv = null;
    private EditText msgEt = null;
    private Button closeBtn = null;
    private Button sendBtn = null;

    @Override
    protected void onCreate(Bundle savedInstanceState) {
        super.onCreate(savedInstanceState);
        setContentView(R.layout.act_bluetooth_chat);

        contentTv = (TextView) findViewById(R.id.act_chat_content_tv);
        msgEt = (EditText) findViewById(R.id.act_chat_msg_et);
        closeBtn = (Button) findViewById(R.id.act_chat_close_btn);
        sendBtn = (Button) findViewById(R.id.act_chat_send_btn);

        closeBtn.setOnClickListener(onBtnClickListener);
        sendBtn.setOnClickListener(onBtnClickListener);
```

```
        bluetoothAdapter = BluetoothAdapter.getDefaultAdapter();

        showDevice();
        //创建服务线程并启动
        serverThread = new ServerThread();
        serverThread.start();
    }

    private OnClickListener onBtnClickListener = new OnClickListener() {
        @Override
        public void onClick(View v) {
            if (v == closeBtn) {
                finish();
            } else if (v == sendBtn) {
                String msg = msgEt.getText().toString();
                if (msg.isEmpty()) {
                    Toast.makeText(BluetoothServiceActivity.this,
                            "内容不能为空",Toast.LENGTH_SHORT).show();
                } else {
                    sendMessageToRemote(msg);
                    msgEt.setText("");
                }
            }
        }
    };
}
```

打开服务器端的页面后，需要请求设备可见，以便客户端能够扫描到服务器端设备，接下来实现 showDevice()方法，代码如下：

```
/**
 * 设置设备可见
 */
private void showDevice() {
    // 向用户发出请求，使当前蓝牙设备可见
    Intent discoverableIntent = new Intent(
            BluetoothAdapter.ACTION_REQUEST_DISCOVERABLE);
    // 设置可见时间为 300 秒
    discoverableIntent.putExtra(
            BluetoothAdapter.EXTRA_DISCOVERABLE_DURATION, 300);
    startActivityForResult(discoverableIntent, 0);
}
```

在上述代码中，通过 startActivityForResult()方法向用户发出请求，使当前蓝牙设备可见，若需要得知是否成功打开可见状态，可以通过重写 onActivityResult()方法的响应码(resultCode)进行判断，如果为“RESULT_OK”，表示同意打开。

接下来需要编写将通信内容显示到页面的代码，考虑到监听客户端的操作是在子线程中进行的，由于子线程不能直接更新 UI 主线程，因此需要 Handler 机制进行处理。代码如下：

```
Handler handler = new Handler() {
        @Override
        public void handleMessage(Message msg) {
                contentTv.append((String) msg.obj + "\n");
        }
};

/**
 * 显示消息到页面
 *
 * @param what
 * @param msgStr
 */
public void showMessageToUI(final String msgStr) {
        Message msg = new Message();
        msg.obj = msgStr;
        handler.sendMessage(msg);
}
```

在上述代码中，创建了 Handler 类，用于将文本显示到通信界面上，在子线程中，只需调用 showMessageToUI()方法，传入要显示的文本即可。该方法会创建 Message 对象，用于封装文本数据，并将其发送到 Handler 中进行处理。

接下来编写服务端线程，用于启动服务器端 Socket 服务，监听客户端的连接请求，以及发送数据。该服务端线程是以内部类的形式编写的，代码如下：

```
/**
 * 服务端线程
 *
 */
private class ServerThread extends Thread {
        public void run() {
                InputStream in = null;
                try {
                        // 监听通信通道
                        serverSocket = bluetoothAdapter
                                .listenUsingRfcommWithServiceRecord("ServiceName",
```

```
                    UUID.fromString(MainActivity.BLUETOOTH_UUID));
            showMessageToUI("系统：服务已启动");

            /* 接受客户端的连接请求 */
            socket = serverSocket.accept();
            // 获取客户端设备的名称
            BluetoothDevice clientDevice = socket.getRemoteDevice();
            remoteName = clientDevice.getName();
            showMessageToUI("系统：客户端(" + remoteName + ")已连接");
            // 等待读取客户端发送的数据
            byte[] buffer = new byte[512];
            in = socket.getInputStream();
            int c = 0;
            while ((c = in.read(buffer)) > 0) {
                String str = new String(buffer, 0, c, "gbk");
                showMessageToUI(remoteName + "：  " + str);
            }
        } catch (IOException e) {
            e.printStackTrace();
            showMessageToUI("系统：服务启动失败！ ");
        } finally {
            try {
                if (in != null) {
                    in.close();
                }
            } catch (IOException e) {
                e.printStackTrace();
            }
        }
    }
};
```

该类主要实现以下内容：

- ✧ 注册监听 Socket 通信通道：需要定义一个 UUID，客户端与服务器建立连接时必须使用相同的 UUID，这里使用的 UUID 以静态常量的方式定义到了 MainActivity 中。
- ✧ accept()方法会等待客户端的连接，一旦连接成功，会返回与客户端通信的 BluetoothSocket 对象。
- ✧ 通过 BluetoothSocket 对象的 getInputStream()方法获取 InputStream 输入流。
- ✧ 之后执行 InputStream 对象的 read()方法，该方法是阻塞的，会一直等待读取客户端发送的数据。

实现发送消息到客户端的功能，代码如下：

```
/**
 * 发送消息到远端设备
 * @param msg
 */
private void sendMessageToRemote(String msg) {
        if (socket == null) {
                Toast.makeText(this, "未连接远程设备", Toast.LENGTH_SHORT).show();
                return;
        }
        try {
                OutputStream os = socket.getOutputStream();
                os.write(msg.getBytes());
                contentTv.append("我：" + msg + "\n");
        } catch (IOException e) {
                e.printStackTrace();
                contentTv.append("我：发送失败\n");
        }
}
```

在该方法中，通过 BluetoothSocket 对象获取 OutputStream 输出流，向远端设备发送文本消息，之后将发送的消息显示到界面上。

至此，服务器端的功能均已实现。当关闭服务器时，需要先关闭连接、释放资源。把这部分代码添加到 onDestroy()生命周期方法中，代码如下：

```
@Override
protected void onDestroy() {
        super.onDestroy();
        if (serverThread != null) {
                serverThread.interrupt();
                serverThread = null;
        }
        try {
                if (socket != null) {
                        socket.close();
                        socket = null;
                }
                if (serverSocket != null) {
                        serverSocket.close();
                        serverSocket = null;
                }
        } catch (IOException e) {
```

```
            e.printStackTrace();
        }
}
```

(3) 实现客户端的代码编写，主要功能有：

- ✧ 创建蓝牙通信通道，连接到服务端。
- ✧ 读取服务端发送的消息。
- ✧ 发送消息到服务端。

创建“BluetoothClientActivity.java”类，该类中的代码将分步骤实现，首先编写基本代码，包括布局文件中控件的引用等，代码如下：

```
public class BluetoothClientActivity extends Activity {
    /** 远端蓝牙设备名称 */
    private String remoteName = null;
    /** 远端蓝牙设备地址 */
    private String remoteAddress = null;

    /** 远端蓝牙设备 */
    private BluetoothDevice remoteDevice = null;
    private BluetoothAdapter bluetoothAdapter = null;

    /** 客户端线程 */
    private ClientThread clientThread = null;
    private BluetoothSocket socket = null;

    private TextView contentTv = null;
    private EditText msgEt = null;
    private Button closeBtn = null;
    private Button sendBtn = null;

    @Override
    protected void onCreate(Bundle savedInstanceState) {
        super.onCreate(savedInstanceState);
        setContentView(R.layout.act_bluetooth_chat);

        contentTv = (TextView) findViewById(R.id.act_chat_content_tv);
        msgEt = (EditText) findViewById(R.id.act_chat_msg_et);
        closeBtn = (Button) findViewById(R.id.act_chat_close_btn);
        sendBtn = (Button) findViewById(R.id.act_chat_send_btn);

        closeBtn.setOnClickListener(onBtnClickListener);
        sendBtn.setOnClickListener(onBtnClickListener);
```

```
            bluetoothAdapter = BluetoothAdapter.getDefaultAdapter();
            // 获取主界面传递的数据
            Intent intent = getIntent();
            remoteAddress = intent.getStringExtra(
                    MainActivity.TAG_REMOTE_ADDRESS);
            remoteName = intent.getStringExtra(MainActivity.TAG_REMOTE_NAME);

            initClient();
        }
        private OnClickListener onBtnClickListener = new OnClickListener() {

            @Override
            public void onClick(View v) {
                if (v == closeBtn) {
                    finish();
                } else if (v == sendBtn) {
                    String msg = msgEt.getText().toString();
                    if (msg.isEmpty()) {
                        Toast.makeText(BluetoothClientActivity.this,
                                "请输入内容",Toast.LENGTH_SHORT).show();
                    } else {
                        // 发送消息到远端设备
                        sendMessageToRemote(msg);
                        msgEt.setText("");
                    }
                }
            }
        };
}
```

上述代码在 onCreate()方法中，涉及两个常量：TAG_REMOTE_ADDRESS 和 TAG_REMOTE_NAME。这两个常量被定义到 MainActivity 类中，分别表示服务端的地址和名称。

接下来需要编写将通信内容显示到页面部分的代码，此部分代码与服务器端代码相同，直接复制即可，复制内容为创建 Handler 对象、showMessageToUI()方法。

之后编写客户端线程，用于与服务端建立连接，监听服务端发送的数据。该客户端线程也是以内部类的形式编写的，代码如下：

```
/**
 * 初始化客户端，与服务器建立连接
 *
 */
```

```
private void initClient() {
        String str = "系统：正在连接 " + remoteName + "(" + remoteAddress + ")";
        contentTv.append(str + "\n");
        // 通过 MAC 地址获取远端蓝牙设备
        remoteDevice = bluetoothAdapter.getRemoteDevice(remoteAddress);
        // 创建客户端线程
        clientThread = new ClientThread();
        clientThread.start();
}
/**
 * 客户端线程
 *
 */
private class ClientThread extends Thread {
        public void run() {
                InputStream in = null;
                try {
                        // 创建通信通道
                        socket = remoteDevice.createRfcommSocketToServiceRecord(
                                UUID.fromString(MainActivity.BLUETOOTH_UUID));
                        // 连接远程设备
                        socket.connect();
                        showMessageToUI("系统：连接成功！");

                        // 等待读取服务器发送的数据
                        byte[] buffer = new byte[512];
                        in = socket.getInputStream();
                        int c = 0;
                        while ((c = in.read(buffer)) > 0) {
                                String str = new String(buffer, 0, c, "gbk");
                                showMessageToUI(remoteName + "： " + str);
                        }
                } catch (IOException e) {
                        showMessageToUI("系统：连接失败！");
                        e.printStackTrace();
                } finally {
                        try {
                                if (in != null) {
                                        in.close();
                                }
```

```
                } catch (IOException e) {
                    e.printStackTrace();
                }
            }
        }
};
```

在上述代码中，initClient()方法用于初始化客户端相关信息，包括提示信息、通过服务端地址获取设备对象、创建并启动客户端线程等。ClientThread 类中通过与服务端相同的 UUID 创建 Socket 通信通道，与服务端进行连接，之后等待读取服务端发送的数据，该部分代码与服务端线程代码类似。

接下来实现发送消息到服务端的代码，该部分代码与服务端发送消息的代码相同，直接复制即可，相关方法为 sendMessageToRemote()。

至此，客户端的功能均已实现，当关闭界面时，需要先关闭连接、释放资源。onDestroy()生命周期方法中代码如下：

```
@Override
protected void onDestroy() {
    super.onDestroy();
    // 断开客户端连接 ，关闭线程
    if (socket != null) {
        try {
            socket.close();
        } catch (IOException e) {
            e.printStackTrace();
        }
        socket = null;
    }
    if (clientThread != null) {
        clientThread.interrupt();
        clientThread = null;
    }
}
```

(4) 继续实现主界面的编程。修改“activity_main.xml”布局文件，代码如下：

```
<LinearLayout xmlns:android="http://schemas.android.com/apk/res/android"
    xmlns:tools="http://schemas.android.com/tools"
    android:layout_width="match_parent"
    android:layout_height="match_parent"
    android:background="#ededed"
    android:orientation="vertical"
    android:padding="10dp" >
```

```
<RelativeLayout
        android:layout_width="fill_parent"
        android:layout_height="wrap_content" >

    <TextView
            android:id="@+id/tv_name"
            android:layout_width="wrap_content"
            android:layout_height="wrap_content"
            android:layout_centerVertical="true"
            android:text="蓝牙名称："
            android:textSize="16sp" />

    <EditText
            android:id="@+id/act_main_btname_et"
            android:layout_width="fill_parent"
            android:layout_height="wrap_content"
            android:layout_centerVertical="true"
            android:layout_toLeftOf="@+id/act_main_change_name_btn"
            android:layout_toRightOf="@+id/tv_name"
            android:hint="本地蓝牙名称" />

    <Button
            android:id="@+id/act_main_change_name_btn"
            android:layout_width="wrap_content"
            android:layout_height="wrap_content"
            android:layout_alignParentRight="true"
            android:layout_centerVertical="true"
            android:text="修改名称" />
</RelativeLayout>

<Button
        android:id="@+id/act_main_find_btn"
        android:layout_width="fill_parent"
        android:layout_height="wrap_content"
        android:text="扫描蓝牙设备" />

    <Button
        android:id="@+id/act_main_service_btn"
        android:layout_width="fill_parent"
        android:layout_height="wrap_content"
```

```
        android:text="作为服务器" />

    <TextView
        android:layout_width="fill_parent"
        android:layout_height="wrap_content"
        android:layout_marginTop="10dp"
        android:background="#dbadff"
        android:padding="5dp"
        android:text="已配对的设备："
        android:textSize="16sp" />

    <ListView
        android:id="@+id/act_main_device_bonded_list"
        android:layout_width="fill_parent"
        android:layout_height="wrap_content"
        android:layout_marginTop="10dp" />

    <TextView
        android:layout_width="fill_parent"
        android:layout_height="wrap_content"
        android:layout_marginTop="10dp"
        android:background="#dbadff"
        android:padding="5dp"
        android:text="查找到新的设备："
        android:textSize="16sp" />

    <ListView
        android:id="@+id/act_main_device_new_list"
        android:layout_width="fill_parent"
        android:layout_height="wrap_content"
        android:layout_marginTop="10dp" />

</LinearLayout>
```

布局中主要有三部分：

- ✧ 第一部分为本地蓝牙设备名称的显示与修改。
- ✧ 第二部分有两个按钮，当点击“扫描蓝牙设备”按钮时，程序开始扫描周围可见的蓝牙设备；当点击“作为服务器”按钮时，可将当前设备作为 Socket 服务器。
- ✧ 第三部分是设备显示区域，列出的设备分为“已配对的设备”和“查找到新的设备”。

(5) 主页面中，是用两个 ListView 控件分别列出不同状态的远端蓝牙设备，因此需要为 ListView 创建一个“适配器”。首先为这个适配器创建 Item 布局文件“item_device_list.xml”，代码如下：

```
<LinearLayout xmlns:android="http://schemas.android.com/apk/res/android"
    xmlns:tools="http://schemas.android.com/tools"
    android:layout_width="match_parent"
    android:layout_height="match_parent"
    android:orientation="vertical"
    android:padding="10dp" >

<RelativeLayout
        android:layout_width="fill_parent"
        android:layout_height="wrap_content" >

    <TextView
            android:id="@+id/item_device_name_tv"
            android:layout_width="wrap_content"
            android:layout_height="wrap_content"
            android:text="蓝牙名称"
            android:textSize="16sp" />

    <TextView
            android:id="@+id/item_device_state_tv"
            android:layout_width="wrap_content"
            android:layout_height="wrap_content"
            android:layout_alignParentRight="true"
            android:text="状态"
            android:textSize="16sp" />
</RelativeLayout>

    <TextView
        android:id="@+id/item_device_mac_tv"
        android:layout_width="wrap_content"
        android:layout_height="wrap_content"
        android:layout_marginLeft="10dp"
        android:text="MAC"
        android:textSize="16sp" />
</LinearLayout>
```

之后创建“BluetoothListAdapter.java”类，并继承“BaseAdapter”，代码如下：

```
package com.yg.ch06_bluetooth_info;
```

```
public class BluetoothListAdapter extends BaseAdapter {
    private Context context = null;
    private List<BluetoothDevice> devices = null;

    public BluetoothListAdapter(Context context,
        List<BluetoothDevice> devices) {
            this.context = context;
            this.devices = devices;
    }

    @Override
    public int getCount() {
            return devices.size();
    }

    @Override
    public Object getItem(int position) {
            return devices.get(position);
    }

    @Override
    public long getItemId(int position) {
            return 0;
    }

    @Override
    public View getView(int position,View convertView,ViewGroup parent){

            BluetoothDevice device = devices.get(position);
            if (convertView == null) {
                    convertView = LayoutInflater.from(context).inflate(
                                    R.layout.item_device_list, null);
            }
            TextView nameTv = (TextView) convertView
                            .findViewById(R.id.item_device_name_tv);
            TextView macTv = (TextView) convertView
                            .findViewById(R.id.item_device_mac_tv);
            TextView stateTv = (TextView) convertView
                            .findViewById(R.id.item_device_state_tv);
```

```
            nameTv.setText(device.getName());
            macTv.setText(device.getAddress());
            stateTv.setText(convertState(device.getBondState()));
            return convertView;
    }

    /**
     * 将数字状态转换为文本
     *
     * @param bondState
     * @return
     */
    private String convertState(int bondState) {
            switch (bondState) {
            case BluetoothDevice.BOND_BONDED:
                    return "已配对";
            case BluetoothDevice.BOND_BONDING:
                    return "配对中";
            case BluetoothDevice.BOND_NONE:
                    return "未配对";
            }
            return "未知状态";
    }
}
```

在上述代码中，通过继承 BaseAdapter 类实现了自定义适配器，主要用于解析 Activity 中传入的蓝牙设备列表(List<BluetoothDevice>devices)，然后把它显示到 ListView 控件中，显示的内容为蓝牙设备的名称、MAC 地址和状态。

(6) 之后实现主界面编程，该类主要实现功能如下：

- ✧ 获取本地蓝牙设备名称，并可修改。
- ✧ 显示已配对的设备列表。
- ✧ 查找并显示周围蓝牙设备。
- ✧ 点击列表中的设备进行连接。

修改“MainActivity.java”类，该类中的代码将分步骤实现，首先编写基本代码，包括布局文件中控件的引用等，代码如下：

```
public class MainActivity extends Activity {
    /** 客户端和服务器通信的 UUID */
    public static final String BLUETOOTH_UUID =
                    "00001101-0000-1000-8000-00805F9B34FB";
    /** 远端设备地址-标签 */
    public static final String TAG_REMOTE_ADDRESS = "remote_address";
```

```
/** 远端设备名称-标签 */
public static final String TAG_REMOTE_NAME = "remote_name";
/** 请求打开蓝牙设备 */
private static final int BLUETOOTH_REQUEST_ENABLE = 1;
/** 蓝牙适配器 */
private BluetoothAdapter bluetoothAdapter = null;

private EditText nameEt = null;
/** 修改蓝牙名称按钮 */
private Button changeNameBtn = null;
/** 查找远端蓝牙设备按钮 */
private Button findBtn = null;
/** 作为服务器按钮 */
private Button serviceBtn = null;
/** 已配对的蓝牙设备列表 */
private ListView bondedDeviceLv = null;
/** 新的蓝牙设备列表 */
private ListView newDeviceLv = null;

/** 进度框 */
private ProgressDialog pDialog = null;

/** 已配对蓝牙设备集合 */
private List<BluetoothDevice> bondedDevices =
        new ArrayList<BluetoothDevice>();
/** 新的蓝牙设备集合 */
private List<BluetoothDevice> newDevices =
        new ArrayList<BluetoothDevice>();

@Override
protected void onCreate(Bundle savedInstanceState) {
        super.onCreate(savedInstanceState);
        setContentView(R.layout.activity_main);

        // 创建 ProgressDialog 进度框，并定义其属性
        pDialog = new ProgressDialog(this);
        pDialog.setCancelable(false);
        pDialog.setMessage("正在扫描蓝牙设备");

        nameEt = (EditText) findViewById(R.id.act_main_btname_et);
```

```
        changeNameBtn=
                (Button) findViewById(R.id.act_main_change_name_btn);
        findBtn = (Button) findViewById(R.id.act_main_find_btn);
        serviceBtn = (Button) findViewById(R.id.act_main_service_btn);
        bondedDeviceLv =
                (ListView)findViewById(R.id.act_main_device_bonded_list);
        newDeviceLv =
                (ListView) findViewById(R.id.act_main_device_new_list);
        bondedDeviceLv.setOnItemClickListener(onListItemClickListener);
        newDeviceLv.setOnItemClickListener(onListItemClickListener);

        changeNameBtn.setOnClickListener(onBtnClickListener);
        findBtn.setOnClickListener(onBtnClickListener);
        serviceBtn.setOnClickListener(onBtnClickListener);

        bluetoothAdapter = BluetoothAdapter.getDefaultAdapter();
        // 检测蓝牙设备是否已开启
        if (bluetoothAdapter.isEnabled()) {
                initBluetooth();
        } else {
                openBluetooth();
        }
}

private OnClickListener onBtnClickListener = new OnClickListener() {
        @Override
        public void onClick(View v) {

                if (v == changeNameBtn) {
                        // 修改本地蓝牙名称
                        String name = nameEt.getText().toString();
                        boolean rst = bluetoothAdapter.setName(name);
                        Toast.makeText(MainActivity.this, rst ?
                                "名称已修改" : "修改失败",
                                        Toast.LENGTH_SHORT).show();
                } else if (v == findBtn) {
                        findBluetooth();
                } else if (v == serviceBtn) {
                        // 打开服务器通信页面
                        Intent intent = new Intent(getApplicationContext(),
```

```
                                BluetoothServiceActivity.class);
                        startActivity(intent);
                    }
                }
            };
}
```

上述代码，对程序中用到的常量进行了定义，实现了按钮的监听事件。开启蓝牙设备通过 openBluetooth()方法来实现。该方法通过向用户发出请求的方式开启蓝牙设备，重写 onActivityResult()方法监听蓝牙是否已启动，代码如下：

```
/**
 * 开启蓝牙
 */
protected void openBluetooth() {
        // 向用户发出请求，开启蓝牙设备
        Intent intent = new Intent(BluetoothAdapter.ACTION_REQUEST_ENABLE);
        startActivityForResult(intent, BLUETOOTH_REQUEST_ENABLE);
}
@Override
protected void onActivityResult(int requestCode, int resultCode,
    Intent data) {
        super.onActivityResult(requestCode, resultCode, data);
        if (requestCode == BLUETOOTH_REQUEST_ENABLE) {
                if (resultCode ==RESULT_OK) {
                Toast.makeText(this,"蓝牙开启成功",Toast.LENGTH_SHORT).show();
                initBluetooth();
                } else {
                Toast.makeText(this,"蓝牙开启失败",Toast.LENGTH_SHORT).show();
                }

        }
}
```

当蓝牙设备已开启后，调用 initBluetooth()方法显示蓝牙名称等信息。代码如下：

```
private void initBluetooth() {
        // 获取本机蓝牙设备名称，显示到 TextView
        String btName = bluetoothAdapter.getName();
        nameEt.setText(btName);
        // 显示历史配对的蓝牙设备
        showBondedBluetooth();
}
```

接下来完成查找远端蓝牙设备编码：首先创建 BroadcastReceiver 广播接收器，用于接收发现蓝牙设备和查找完成的广播；然后注册这个广播接收器；最后执行查找远端蓝牙设备请求。代码如下：

```
/**
 * 查找远端蓝牙设备
 */
protected void findBluetooth() {
        pDialog.show();
        registReceiver();
        // 开始查找蓝牙设备
        bluetoothAdapter.startDiscovery();
}

/**
 * 注册监听
 */
private void registReceiver() {
        IntentFilter fileter = new IntentFilter();
        fileter.addAction(BluetoothDevice.ACTION_FOUND);
        fileter.addAction(BluetoothAdapter.ACTION_DISCOVERY_FINISHED);
        registerReceiver(bluetoothReceiver, fileter);
}

private final BroadcastReceiver bluetoothReceiver =
    new BroadcastReceiver() {
        @Override
        public void onReceive(Context context, Intent intent) {
                String action = intent.getAction();
                if (BluetoothDevice.ACTION_FOUND.equals(action)) {
                        BluetoothDevice device =
                        intent.getParcelableExtra(BluetoothDevice.EXTRA_DEVICE);
                        if (device.getBondState()!= BluetoothDevice.BOND_BONDED) {
                                // 去除重复添加的设备
                                if (!newDevices.contains(device)) {
                                   newDevices.add(device);
                                }
                        }
                } else if (BluetoothAdapter.ACTION_DISCOVERY_FINISHED
                                .equals(action)) {
                        pDialog.dismiss();
                        unregisterReceiver(bluetoothReceiver);
```

```
                    //扫描完毕，显示查找到的蓝牙设备
                    showFoundDevices();
                }
            }
};
```

当完成查找远端蓝牙设备操作后，通过 showFoundDevices()方法将查找到的蓝牙设备绑定到 ListView，代码如下：

```
/**
 * 显示已经找到的蓝牙设备
 */
protected void showFoundDevices() {
        BluetoothListAdapter bondedAdapter = new BluetoothListAdapter(this,
                        newDevices);
        newDeviceLv.setAdapter(bondedAdapter);
}
```

至此，扫描远端蓝牙操作已经完成。接下来编写用于显示已配对设备列表的代码如下：

```
/**
 * 显示已配对蓝牙设备
 */
protected void showBondedBluetooth() {
        Set<BluetoothDevice> deviceSet = bluetoothAdapter.getBondedDevices();
        if (deviceSet.size() == 0) {
                Toast.makeText(this, "没有已配对设备", Toast.LENGTH_SHORT).show();
                return;
        }
        bondedDevices = new ArrayList<BluetoothDevice>(deviceSet);
        BluetoothListAdapter bondedAdapter = new BluetoothListAdapter(this,
                        bondedDevices);
        bondedDeviceLv.setAdapter(bondedAdapter);
}
```

最后要实现点击列表中的设备，进行连接，对 ListView 添加 OnItemClickListener 监听器。本示例中有已配对设备列表和新的设备列表两个 ListView，使用同一个监听器即可，代码如下：

```
private OnItemClickListener onListItemClickListener =
    new OnItemClickListener() {
        @Override
        public void onItemClick(AdapterView<?> parent, View view,
            int position,long id) {
                BluetoothDevice device = null;
```

```
            // 判断点击的 ListView，获取相应的设备对象
            if (parent == bondedDeviceLv) {
                device = bondedDevices.get(position);
            } else if (parent == newDeviceLv) {
                device = newDevices.get(position);
            }
            // 打开客户端通信页面
            Intent intent = new Intent(getApplicationContext(),
                        BluetoothClientActivity.class);
            intent.putExtra(TAG_REMOTE_NAME, device.getName());
            intent.putExtra(TAG_REMOTE_ADDRESS, device.getAddress());
            startActivity(intent);
    }
};
```

在代码中，首先对点击了哪个 ListView 进行判断，之后获取相应的蓝牙设备对象，创建 Intent，并将设备的名称和地址添加到 Intent 中，最后启动客户端页面。

至此，“MainActivity.java”类中的代码已完成编写。

(7) 在“AndroidManifest.xml”文件中进行相关配置，注册 Activity 类的代码不再赘述。注册蓝牙操作相关权限，代码如下：

```
<uses-permission android:name="android.permission.BLUETOOTH" />
<uses-permission android:name="android.permission.BLUETOOTH_ADMIN" />
```

(8) 分别在两个手机中运行本程序，并进行通信，其中一部做服务端，另一部做客户端。扫描设备完成界面如图 6-4 所示，客户端与服务器通信界面如图 6-5 所示。

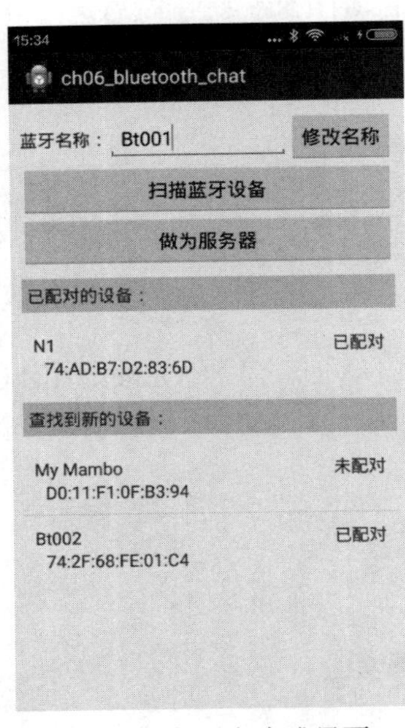

图 6-4　扫描设备完成界面

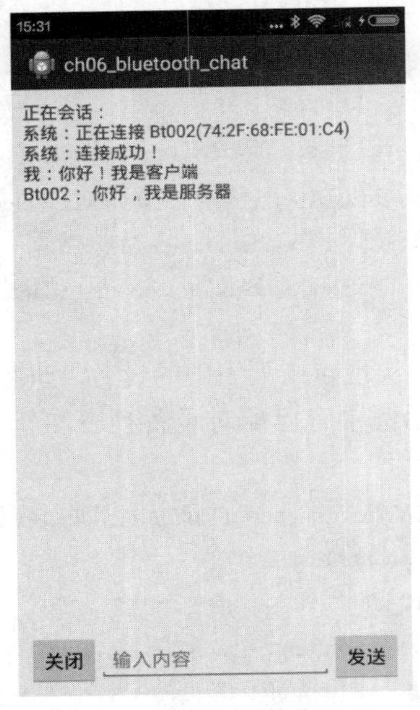

图 6-5　与服务器通信界面

6.2.3　BLE 技术概述

BLE 是“Bluetooth Low Energy”的缩写，叫作低功耗蓝牙设备。相对于传统蓝牙技术，该技术具有更低的功耗，一节纽扣电池足可以使其工作数年之久。被广泛应用于蓝牙耳机、蓝牙音箱以及物联网的实现中。

本小节示例中用到的 BLE 技术有以下相关类。

1．BluetoothGattCharacteristic

该类是构造 BluetoothGattService 的基本元素，包含 1 个 Value 值和 0 或多个描述信息，该对象通过一个 UUID 唯一标识。常用方法如表 6-12 所示。

表 6-12　BluetoothGattCharacteristic 类常用方法

方 法 名	描　述
addDescriptor(BluetoothGattDescriptor descriptor)	添加一个描述对象
getDescriptor(UUID uuid)	通过传入的 UUID 获取指定描述对象
getDescriptors()	获取描述对象列表
getService()	获取 characteristic 所属的服务对象
getPermissions()	获取 characteristic 的权限
getProperties()	获取 characteristic 的属性
getUuid()	获取 UUID
getValue()	获取 characteristic 存储的值
setValue(String value)	设置 characteristic 存储的值

2．BluetoothGattService

该类用于描述蓝牙 GATT 协议服务，是 BluetoothGattCharacteristic 的集合，该对象通过一个 UUID 唯一标识。常用方法如表 6-13 所示。

表 6-13　BluetoothGattService 类常用方法

方 法 名	描　述
addCharacteristic(BluetoothGattCharacteristic characteristic)	添加 BluetoothGattCharacteristic
addService(BluetoothGattService service)	添加 BluetoothGattService
getCharacteristic(UUID uuid)	通过传入的 UUID 获取指定的 BluetoothGattCharacteristic 对象
getCharacteristics()	获取 BluetoothGattCharacteristic 列表
getType()	获取服务类型(主/从设备)
getUuid()	获取 UUID

3．BluetoothGatt

该类是蓝牙 GATT 协议描述对象，用于维护与远端设备的连接状态与远端设备通信等。常用方法如表 6-14 所示。

表 6-14　BluetoothGatt 类常用方法

方　法　名	描　　述
connect()	连接到远端设备
close()	关闭蓝牙 GATT 协议客户端
disconnect()	断开建立的连接
discoverServices()	开启发现远端设备的服务以及特征和描述符等
getDevice()	获取远端蓝牙设备
getServices()	获取 GATT 远端设备提供的服务列表
setCharacteristicNotification (BluetoothGattCharacteristic characteristic, boolean enable)	设置当指定 characteristic 值发生变化时，是否发出通知
writeCharacteristic (BluetoothGattCharacteristic characteristic)	写入一个 characteristic 到远端设备
readRemoteRssi()	获取远端设备的 RSSI

4．BluetoothManager

该类用于管理蓝牙的相关功能，例如获取蓝牙适配器、已连接的蓝牙列表、连接状态等。常用方法如表 6-15 所示。

表 6-15　BluetoothManager 类常用方法

方　法　名	描　　述
getAdapter()	获取默认的蓝牙适配器
getConnectedDevices(int profile)	获取已连接的设备列表，参数“profile”表示设备类型： BluetoothGatt.GATT 从设备 BluetoothGatt.GATT _SERVER 主设备
getConnectionState(BluetoothDevice device, int profile)	获取设备连接状态，返回的状态常量值被定义在 BluetoothGatt 类中，包括： STATE_CONNECTED 已连接 STATE_CONNECTING 正在连接 STATE_DISCONNECTED 已断开 STATE_DISCONNECTING 正在断开

5．BluetoothGattCallback

该类是蓝牙 GATT 协议回调。创建 GATT 协议连接时，需要传入该对象，用于实现通信相关监听。该类中常用回调方法如表 6-16 所示。

表 6-16　BluetoothGattCallback 类常用方法

方　法　名	描　　述
onConnectionStateChange(BluetoothGatt gatt, int status, int newState)	连接状态改变时发生回调
onServicesDiscovered(BluetoothGatt gatt, int status)	发现新的服务时发生回调
onCharacteristicChanged(BluetoothGatt gatt, BluetoothGattCharacteristic characteristic)	characteristic 发生变化时发生回调

6.2.4 通过 BLE 技术与设备通信

本小节将利用 BLE 技术与开发板(设备)进行通信，模拟物联网的应用。示例中用的是串口蓝牙 4.0 模块，如图 6-6 所示，将此模块插入开发板相应接口即可。

图 6-6　串口蓝牙 4.0 模块

需要注意的是：在使用 BLE 技术时，设备间的连接和通信必须具有相同的 UUID。因此，在开发应用程序之前，首先需要了解蓝牙模块的 UUID，不同的蓝牙模块生产厂商可能会设置不同的 UUID，这需要查阅模块对应的技术手册来获取。本示例蓝牙模块型号是“DX-BT05”，支持的服务是“Central & Peripheral UUID FFE0, FFE1”。从支持的服务信息可得知：UUID 为“FFE0”和“FFE1”。通常第一个 UUID 指的是 BluetoothGattService 对象的 UUID，在该对象中，可以通过第二个 UUID 获取用于通信的 BluetoothGatt Characteristic 对象。本示例中的 UUID 为：

✧ BluetoothGattService UUID：0000ffe0-0000-1000-8000-00805f9b34fb。

✧ BluetoothGattCharacteristic UUID：0000ffe1-0000-1000-8000-00805f9b34fb。

1．通信流程

本示例中，手机端与设备通信的主从关系为：手机端为“主”，设备为“从”。主要通信流程如下：

(1) 扫描支持 BLE 技术的远端蓝牙设备，扫描设备代码如下：

```
bluetoothAdapter.startLeScan(bleScanCallback);
```

“bleScanCallback”为回调对象，当扫描到远端设备时，会执行该对象中的 onLeScan()回调方法，将设备添加到列表中保存。

(2) 点击扫描结果列表，与相应的设备进行连接，代码如下：

```
BluetoothGattbluetoothGatt = device.connectGatt(this, false, bluetoothGattCallback);
```

当连接到设备后，会返回 BluetoothGatt 对象。“bluetoothGattCallback”为回调对象，当连接状态改变时，会执行该对象中的 onConnectionStateChange()回调方法，判断是否连接成功。

(3) 成功连接 BLE 设备后，开始获取(发现)该设备的所有 GATT 服务(BluetoothGattService)，发现设备服务代码如下：

```
bluetoothGatt.discoverServices();
```

(4) 当获取到新的 GATT 服务，会调用“bluetoothGattCallback”回调对象中的 onServicesDiscovered()方法。在此方法中遍历发现的服务，通过指定的 BluetoothGattService 对象的 UUID 找到对应的服务，然后通过指定的 BluetoothGattCharacteristic 对象的 UUID，获取该服务中对应的 BluetoothGattCharacteristic 对象。

在 onServicesDiscovered()回调方法中获取服务列表的代码如下：

```
List<BluetoothGattService> gattServices = bluetoothGatt.getServices();
```

通过指定 UUID 获取服务中对应的 BluetoothGattCharacteristic 对象代码如下：

```
UUID uuid = UUID.fromString(UUID_KEY_CHARACT);
BluetoothGattCharacteristicgattCharacteristic = gattService.getCharacteristic(uuid);
```

(5) 成功获取到所需要的 BluetoothGattCharacteristic 对象之后，通过该对象可以进行手机端与设备的通信。当设备向手机发送数据时，“bluetoothGattCallback”对象中的onCharacteristicChanged()会被触发，可以通过该方法中的“characteristic”对象获取设备发送的数据，代码如下：

```
byte[] msgByte = characteristic.getValue();
String msgStr=new String(msg);
```

手机端向设备发送数据，代码如下：

```
gattCharacteristic.setValue("Hello!");
boolean rst = bluetoothGatt.writeCharacteristic(gattCharacteristic);
```

2．通信示例

下述示例用于实现：利用 BLE 技术实现与 6.2.2 小节相同的功能。要求如下：

- ✧ 能够扫描周围 BLE 蓝牙设备；
- ✧ 点击设备打开设备控制界面进行通信；
- ✧ 使用与 6.2.2 小节相同的通信协议。

(1) 创建项目“ch06_BLE_IoT”，首先创建主界面中用于显示设备列表的 Adapter 适配器。创建适配器用到的 Item 布局文件“item_device_list.xml”，代码如下：

```
<LinearLayout xmlns:android="http://schemas.android.com/apk/res/android"
    xmlns:tools="http://schemas.android.com/tools"
    android:layout_width="match_parent"
    android:layout_height="match_parent"
    android:orientation="vertical"
    android:padding="10dp" >
<TextView
        android:id="@+id/item_device_name_tv"
        android:layout_width="wrap_content"
        android:layout_height="wrap_content"
        android:text="蓝牙名称"
        android:textSize="16sp" />
<TextView
        android:id="@+id/item_device_mac_tv"
        android:layout_width="wrap_content"
        android:layout_height="wrap_content"
        android:layout_marginLeft="10dp"
        android:text="MAC"
```

```
        android:textSize="16sp" />
</LinearLayout>
```

(2) 创建适配器“LBEDeviceListAdapter.java”类，代码如下：

```
public class LBEDeviceListAdapter extends BaseAdapter {
        private Context context = null;
        private List<BluetoothDevice> devices = null;

        public LBEDeviceListAdapter(Context context,
        List<BluetoothDevice> devices) {
                this.context = context;
                this.devices = devices;
        }
        @Override
        public int getCount() {
                return devices.size();
        }
        @Override
        public Object getItem(int position) {
                return devices.get(position);
        }
        @Override
        public long getItemId(int position) {
                return 0;
        }
        @Override
        public View getView(int position, View convertView,
        ViewGroup parent) {

                BluetoothDevice device = devices.get(position);
                if (convertView == null) {
                        convertView = LayoutInflater.from(context).inflate(
                                        R.layout.item_device_list, null);
                }
                TextView nameTv = (TextView) convertView
                                .findViewById(R.id.item_device_name_tv);
                TextView macTv = (TextView) convertView
                                .findViewById(R.id.item_device_mac_tv);

                nameTv.setText(device.getName());
                macTv.setText(device.getAddress());
```

```
            return convertView;
        }
}
```

(3) 完成主界面的编写，主要功能是扫描周围蓝牙设备。点击设备后，打开设备控制界面。修改“activity_main.xml”布局文件，代码如下：

```
<LinearLayout xmlns:android="http://schemas.android.com/apk/res/android"
    xmlns:tools="http://schemas.android.com/tools"
    android:layout_width="match_parent"
    android:layout_height="match_parent"
    android:background="#ededed"
    android:orientation="vertical"
    android:padding="10dp" >

    <Button
        android:id="@+id/act_main_find_btn"
        android:layout_width="fill_parent"
        android:layout_height="wrap_content"
        android:text="扫描蓝牙设备" />

    <ListView
        android:id="@+id/act_main_lv"
        android:layout_width="fill_parent"
        android:layout_height="wrap_content"
        android:layout_marginTop="10dp" />
</LinearLayout>
```

(4) 修改“MainActivity.java”类，首先编写基础代码，包括控件引用、监听等，代码如下：

```
public class MainActivity extends Activity {

        private LBEDeviceListAdapter deviceListAdapter;
        private BluetoothAdapter bluetoothAdapter;
        private List<BluetoothDevice> devices =
                new ArrayList<BluetoothDevice>();

        private Button scanBtn = null;
        private ListView listView = null;

        @Override
        public void onCreate(Bundle savedInstanceState) {
                super.onCreate(savedInstanceState);
```

```
        setContentView(R.layout.activity_main);

        scanBtn = (Button) findViewById(R.id.act_main_find_btn);
        listView = (ListView) findViewById(R.id.act_main_lv);

        scanBtn.setOnClickListener(new OnClickListener() {
            @Override
            public void onClick(View arg0) {
                scanLeDevice(true);
            }
        });

        // 初始化蓝牙相关操作
        initBlutooth();

        // 初始化设备列表
        initDeviceList();
    }

    /**
     * 初始化蓝牙相关操作
     */
    private void initBlutooth() {
        final BluetoothManager bluetoothManager =
         (BluetoothManager) getSystemService(Context.BLUETOOTH_SERVICE);
        bluetoothAdapter = bluetoothManager.getAdapter();

        // 如果蓝牙不可用，则强制开启
        if (!bluetoothAdapter.isEnabled()) {
            bluetoothAdapter.enable();
        }
    }

    /**
     * 初始化设备列表
     */
    private void initDeviceList() {
        deviceListAdapter = new LBEDeviceListAdapter(this, devices);
        listView.setAdapter(deviceListAdapter);
        listView.setOnItemClickListener(new OnItemClickListener() {
```

```
                @Override
                public void onItemClick(AdapterView<?> arg0, View arg1,
                        int position, long arg3) {
                    final BluetoothDevice device = devices.get(position);
                    Intent intent = new Intent(MainActivity.this,
                            ControlActivity.class);
                    intent.putExtra("device_address",
                        device.getAddress());
                    startActivity(intent);


                }
            });
        }
}
```

接下来添加扫描设备方法，代码如下：

```
private void scanLeDevice(final boolean enable) {
        bluetoothAdapter.startLeScan(bleScanCallback);
        // 10 秒后停止扫描
        new Handler().postDelayed(new Runnable() {
                @Override
                public void run() {
                        bluetoothAdapter.stopLeScan(bleScanCallback);
                }
        }, 1000 * 10);
}
```

在上述代码中，执行扫描蓝牙操作，通过 Handler. postDelayed()方法控制在 10 秒后停止扫描。调用 startLeScan()方法需要传入回调对象，用于发现设备后进行回调操作，通常使用内部类的形式实现，回调类代码如下：

```
private BluetoothAdapter.LeScanCallback bleScanCallback = new
    BluetoothAdapter.LeScanCallback() {


        @Override
        public void onLeScan(final BluetoothDevice device, int rssi,
        byte[] scanRecord) {
                devices.add(device);
                deviceListAdapter.notifyDataSetChanged();
        }
};
```

上述代码将扫描到的蓝牙设备显示到列表中。至此，主界面功能已完成。

(5) 实现设备控制界面，该界面用到的布局文件与 6.2.2 小节中的“act_control_

layout.xml”相同，此处不再赘述。

(6) 创建“ControlActivity.java”类，该类为程序核心类，代码将分步骤实现。首先编写基本代码，包括布局文件中控件的引用等，代码如下：

```
public class ControlActivity extends Activity {
    /** 系统消息 */
    private final static int TAG_WHAT_SYS = 1;
    /** 设备发送来的消息 */
    private final static int TAG_WHAT_DEVICE = 2;

    /** Socket 通信对象 */
    private Socket socket;
    /** 手机端通信线程 */
    private SocketThread socketThread;

    private ScrollView scrollView = null;
    /** 环境信息 */
    private TextView envirInfoTv = null;
    /** 事件信息 */
    private TextView eventInfoTv = null;
    /** 发送屏显 */
    private Button screenBtn = null;
    /** 蜂鸣器报警 */
    private Button buzzBtn = null;
    /** 控制四个 LED 灯 */
    private Switch led1Sw = null;
    private Switch led2Sw = null;
    private Switch led3Sw = null;
    private Switch led4Sw = null;

    @Override
    protected void onCreate(Bundle savedInstanceState) {
        super.onCreate(savedInstanceState);
        setContentView(R.layout.act_control_layout);

        //代码同 6.2.2 小节示例 onCreate()部分
    }
}
```

在上述代码中，由于与 6.2.2 小节示例中用的是相同的布局，所以 onCreate()方法中的代码与之完全相同，此处不再赘述。

接下来建立与设备通信的连接和连接用到的回调对象。回调对象是一个内部类对象，

代码如下：

```
/**
 * 连接到设备
 */
private void initConn() {
        String str = "系统：正在尝试连接设备";
        eventInfoTv.append(str + "\n");

        String address = getIntent().getStringExtra("device_address");
        BluetoothManager bluetoothManager =
                (BluetoothManager) getSystemService(Context.BLUETOOTH_SERVICE);
        bluetoothAdapter = bluetoothManager.getAdapter();
        BluetoothDevice device = bluetoothAdapter.getRemoteDevice(address);
        bluetoothGatt = device.connectGatt(this, false,
                bluetoothGattCallback);
}
private final BluetoothGattCallback bluetoothGattCallback =
      new BluetoothGattCallback() {
        // 当连接状态发生改变
        @Override
        public void onConnectionStateChange(BluetoothGatt gatt, int status,
        int newState) {
                if (newState == BluetoothProfile.STATE_CONNECTED) {
                        // 蓝牙设备已经连接
                        sendMsgToHandler(TAG_WHAT_SYS, "系统：设备已连接");
                        sendMsgToHandler(TAG_WHAT_SYS, "系统：正在查询可用通信服务");
                        bluetoothGatt.discoverServices();
                } else if (newState == BluetoothProfile.STATE_DISCONNECTED) {
                        sendMsgToHandler(TAG_WHAT_SYS, "系统：连接已断开");
                }
        }

        // 发现服务端
        @Override
        public void onServicesDiscovered(BluetoothGatt gatt, int status) {
                if (status == BluetoothGatt.GATT_SUCCESS) {
                        List<BluetoothGattService> gattServices = bluetoothGatt
                                .getServices();
                        for (BluetoothGattService gattService : gattServices) {
                        // 找到 UUID 与 UUID_KEY_SERVER 匹配的 BluetoothGattService
```

```
                    if (gattService.getUuid().toString()
                            .equals(UUID_KEY_SERVER)) {
                    // 通过 UUID_KEY_CHARACT 找到可以与蓝牙模块通信的 Characteristic
                        UUID uuid = UUID.fromString(UUID_KEY_CHARACT);
                        gattCharacteristic = gattService
                                        .getCharacteristic(uuid);
                        if (gattCharacteristic != null) {
                            bluetoothGatt.setCharacteristicNotification(
                                    gattCharacteristic, true);
                                    sendMsgToHandler(TAG_WHAT_SYS,
                                            "系统：通信服务已建立");
                                    return;
                            }
                            break;
                        }
                    }
                    sendMsgToHandler(TAG_WHAT_SYS, "系统：无可用通信服务");
            }
        }

        // Characteristic 对象发生改变
        @Override
        public void onCharacteristicChanged(BluetoothGatt gatt,
                    BluetoothGattCharacteristic characteristic) {
            // 获取 Value 值，发送到 Handler 进行处理
            byte[] msg = characteristic.getValue();
            sendMsgToHandler(TAG_WHAT_DEVICE, new String(msg));
        }
};
```

在上述代码中，“bluetoothGattCallback”回调对象用于监听连接状态、发现新的服务端、监听 Characteristic 对象的改变，也是蓝牙 BLE 通信的核心。当监听到设备发来的数据时，通过 sendMsgToHandler(TAG_WHAT_DEVICE, msg)方法将数据发送到 Handler 进行处理。处理设备发来的数据流程代码如下：

```
/**
 * 发送消息到 Handler
 *
 * @param what
 * @param msgStr
 */
public void sendMsgToHandler(int what, final String msgStr) {
```

```
        //代码同 6.2.2 小节示例，此处略
}

Handler handler = new Handler() {
        @Override
        public void handleMessage(Message msg) {
                //代码同 6.2.2 小节示例，此处略
        }
};

/**
 * 处理设备发送的消息
 *
 * @param msg
 */
private void handleMsgForDevice(String msgStr) {
        //代码同 6.2.2 小节示例，此处略
        scrollToLast();
}
```

上述代码与 6.2.2 小节示例中对应的代码完全相同，此处不再赘述。添加 scrollToLast() 方法，同样与 6.2.2 小节示例相同：

```
private void scrollToLast() {
        //代码同 6.2.2 小节示例，此处略
}
```

至此，单向监听设备发送信息的功能已实现，接下来实现向设备发送指令，控制设备模块的功能。

Button 按钮、Switch 开关控件的事件监听及相关方法的实现代码也与 6.2.2 小节示例代码相同，如下所示：

```
/** 按钮点击事件 */
private OnClickListener onBtnClickListener = new OnClickListener() {
        @Override
        public void onClick(View v) {
                //代码同 6.2.2 小节示例，此处略
        }
};
/** Switch 开关改变事件 */
private OnCheckedChangeListener onSwitchChangeListener =
     new OnCheckedChangeListener() {
        @Override
        public void onCheckedChanged(CompoundButton v, boolean checked) {
                //代码同 6.2.2 小节示例，此处略
```

```
        }
};
/**
 * 显示发送屏显消息对话框
 */
protected void showSendScreenMsgDialog() {
        //代码同 6.2.2 小节示例，此处略}

/**
 * 发送蜂鸣器报警
 */
protected void sendBuzzer() {
        //代码同 6.2.2 小节示例，此处略
}
```

接下来实现 buildCmdDataToDevice()方法，该方法在功能方面与 6.2.2 小节示例相应部分相同，但具体实现的代码不同，代码如下：

```
/**
 * 构建指令并发送指令到设备
 *
 * @param event
 * @param param
 */
private void buildCmdDataToDevice(String event, String param) {
        String cmd = event + ":" + param;
        // 设置数据内容
        gattCharacteristic.setValue(cmd);
        // 往蓝牙模块写入数据
        boolean rst = bluetoothGatt.writeCharacteristic(gattCharacteristic);
        if (rst) {
                eventInfoTv.append("发送 " + event + " 指令成功\n");
        } else {
                eventInfoTv.append("发送 " + event + " 指令失败\n");
        }
        scrollToLast();
}
```

在上述代码中，通过 BluetoothGatt.writeCharacteristic()方法，将指令发送到设备上。

最后，在关闭程序之前需要先断开与设备的连接，释放资源，代码如下：

```
@Override
protected void onDestroy() {
        if (bluetoothGatt != null) {
```

```
            if (bluetoothGatt.connect()) {
                    bluetoothGatt.disconnect();
            }
        }
        super.onDestroy();
}
```

至此，设备控制界面已完成。

(7) 在“AndroidManifest.xml”配置文件中注册“ControlActivity.java”类，代码略。

(8) 在“AndroidManifest.xml”配置文件中声明必要的权限，代码如下：

```
<uses-permission android:name="android.permission.INTERNET" />
<uses-permission android:name="android.permission.ACCESS_WIFI_STATE" />
<uses-permission android:name="android.permission.ACCESS_NETWORK_STATE" />
```

(9) 运行程序进行测试，手机端效果与 6.2.2 小节示例相同(图 6-5)。

本 章 小 结

(1) Wi-Fi 是一项基于 IEEE 802.11 标准的无线网络连接技术。

(2) WifiManager 类用于管理手机中 Wi-Fi 设备的开/关、连接网络、配置信息等操作。

(3) Bluetooth 是一种无线技术标准，工作在 2.4 GHz 频段，理论传输距离为 10 米。

(4) 蓝牙 4.0 有两个分支：传统蓝牙 4.0 技术和 BLE4.0 技术。

(5) BluetoothDevice 类用于描述一个远端的蓝牙设备，该对象可以创建一个连接、获取基本信息和连接状态等。

(6) 设置蓝牙可见的默认时间为 120 秒，可以通过添加附加值的方式改变可见时间，附加值是 BluetoothAdapter.EXTRA_DISCOVERABLE_DURATION。

(7) 两个设备之间建立连接时，UUID 必须相同。

本 章 练 习

(1) 当系统监测到 Wi-Fi 设备状态发生变化后，会发送下列哪一个 Action 广播通知？

(A) NETWORK_STATE_CHANGED_ACTION

(B) RSSI_CHANGED_ACTION

(C) SUPPLICANT_CONNECTION_CHANGE_ACTION

(D) WIFI_STATE_CHANGED_ACTION

(2) 开启设备蓝牙的方式有两种，分别是_________和_________。

(3) _______________类用于管理本地蓝牙设备，例如打开或关闭蓝牙、启用设备发现以及获取蓝牙设备状态等。

(4) 尝试通过蓝牙通信技术在两个手机设备之间传输图片。

(5) 简述蓝牙 4.0 相对于之前蓝牙版本的优点。

第7章　NFC

本章目标

- 理解常见 NFC 数据格式
- 理解标签调度系统
- 掌握常见 NFC 数据格式的读写操作

本章主要讲解 NFC(近场通信技术)。该技术是近几年逐渐被广泛使用的技术，之前常被用于公交刷卡、门禁卡等，目前已经实现通过手机端 APP 为公交卡充值等各种定制化的功能。

7.1　NFC 概述

NFC(近场通信技术)是一种非接触式的短距离通信技术，通常有效距离在 4 cm 之内。这项技术是由 RFID(非接触式射频识别技术)和互联互通技术整合演变而来的，为了更好地推动 NFC 技术的发展与普及，业界创建了一个非营利性的组织——NFC Forum。该组织主要为了促进 NFC 技术的实施与标准化，确保设备和服务之间协同工作，目前，在全球已拥有数百个成员。如今，NFC 技术已与人们的生活息息相关，例如：公交打卡、门禁系统、银行卡闪付系统等。

7.1.1　RFID 射频识别技术

在讲解 NFC 之前，首先需要了解 RFID 射频识别技术。射频识别技术以多种形式出现，如卡扣、贴纸等。贴纸形式的射频识别技术常贴在商场货架上比较贵重商品的外包装或内包装中，主要用于商品的防盗或库存跟踪。此类贴纸式标签大小通常为 5 cm 之内见方，也有圆形样式，常见标签正反面如图 7-1 所示。

图 7-1　常见标签正反面

观察标签可以发现，背面缠绕着许多圈扁平的金属线，这条线圈是标签的天线，线圈末端连接着一个金属块。不同厂家生产的标签，线路布局可能有所不同，这些金属块都是用硅制成的小型集成电路，用于存储少量的逻辑与认证数据，标签可以通过天线将数据发送到读卡器等设备上。

RFID 标签主要分为主动型、被动型，以及两者的结合型。主动型标签具有内置的电池，相对于被动型标签，通信距离更长，可达 10 m 以上；相反的，被动型标签不需要内置电池供电，相对通信距离更短，价格较低，更为小巧，只要保持电路状态良好即可进行通信。

被动型标签在被读卡器等设备扫描时激活，读卡器设备发射短程射频信号，射频信号中的无线电波会使标签中的线圈产生震荡，这是肉眼观察不到的，并将射频信号转换为电力为标签自身供电，使其与读卡器设备通信。

目前，部分手机配备了 NFC 读卡器设备，为了不破坏手机的外观，手机中的 NFC 读卡器天线也被做成了扁平的线圈，类似于标签贴纸的天线。通常，将天线放置到手机的后盖内侧，或者是直接设计到电池的外侧。对于此类设计，在更换新的电池时，需要确保新电池也有 NDF 读卡器天线，否则 NFC 读卡器设备将无法正常工作。中国移动 N1 手机的 NFC 天线如图 7-2 所示。

图 7-2　中国移动 N1 手机的 NFC 天线

7.1.2　NFC 工作模式

NFC 主要支持三种工作模式，分别是读卡器模式、仿真卡模式和点对点模式。

读卡器模式主要通过 NFC 读卡器设备(例如支持 NFC 功能的手机)对带有 NFC 芯片的标签、名片、钥匙扣、报纸等介质读取或写入数据。NFC 贴纸标签如图 7-3 所示。接下来的小节示例中将使用此标签进行编程通信。

图 7-3　NFC 贴纸标签

仿真卡模式可以将支持 NFC 功能的手机等设备作为被动的电子标签(IC 卡)来使用，

例如饭卡刷卡、门禁卡、公交卡以及银行卡支付等。使用时，只需将手机靠近 NFC 读卡器设备，之后进行复杂的安全验证实现功能。

点对点模式可用于两个不同 NFC 设备(如手机)之间进行数据的交换。NFC 通信速度通常高于红外、蓝牙等方式，但是有效距离只能在 4 cm 之内。在 Android4.2 系统之后，当 NFC 设备连接成功后，将自动通过蓝牙技术进行数据的传输，这样一来，可以有效解决传输距离问题，而且不需要开发者通过编码启动蓝牙通信功能，这时系统可自动启用，对于用户是完全透明的，这项技术就是 Android Beam 技术。

7.2 数据格式

由于 NFC 标签生产厂商的不同以及对标签需求的不同，市面上存在着支持不同协议的标签。Android 中，“android.nfc.tech”包下定义了常见的 NFC 标签技术类，如表 7-1 所示。

表 7-1 Android 中 NFC 相关标签类

类　名	描　述
TagTechnology	所有标签技术类必须实现的接口
NfcA	提供了对 NFC-A(ISO 14443-3A)的属性访问和 I / O 操作
NfcB	提供了对 NFC-A NFC-B (ISO 14443-3B) 的属性访问和 I / O 操作
NfcF	提供了对 NFC-A NFC-F (JIS 6319-4) 的属性访问和 I / O 操作
NfcV	提供了对 NFC-A NFC-V (ISO 15693) 的属性访问和 I / O 操作
IsoDep	提供了对 NFC-A ISO-DEP (ISO 14443-4) 的属性访问和 I / O 操作
Ndef	提供了对 NFC 标签已被格式化为 NDEF 的数据的操作访问
NdefFormatable	提供可能被格式化为 NDEF 的 formattable 的标签，还没有被格式化为 NDEF 格式数据
MifareClassic	如果设备支持 MIFARE，可提供对 MIFARE Classic 的属性访问和 I / O 操作
MifareUltralight	如果设备支持 MIFARE，可提供对 MIFARE 的超轻属性访问和 I / O 操作

由于 Android API 全面的支持该数据格式的操作，因此本书针对不同的 NFC 标签技术做了分类，分为支持 NDEF 数据格式和非 NDEF 数据格式两类标签。两者的主要区别在于：前者有固定的标准来执行数据的操作，比如存储纯文本、URL 等特定数据，加之 Android API 的支持，开发者在操作 NDEF 格式标签时会比较方便灵活。接下来将对这两类技术进行具体的分析。

1. NDEF 数据格式

数据交换格式(NFC Data Exchange Format，NDEF)是 NFC Forum 组织约定的一种 NFC 数据格式。它是轻量级的紧凑的二进制格式，可带有 URL、vCard 和 NFC 定义的各种数据类型。

NDEF 数据被封装到一个消息对象(NdefMessage)中，该消息包含了一个或多个记录(NdefRecord)，每个记录包含了一个有效的负载(数据)，除了有效负载，每个记录需要为

有效负载定义元数据，例如数据的类型和长度等：

- ✧ 有效负载的长度：使用一个无符号整数来表示有效负载的字节长度(一个字节为 8 位)。
- ✧ 有效负载的类型：开发者可以为程序声明不同的类型，例如 URI、电子名片(text/x-vCard)、纯文本(text/plain-text)、MIME 媒介等。另外，Android API 定义了 TNF(Type Name Format)值对有效负载进行了分类。这些 TNF 常量被定义到了 NdefRecord 类中，例如："TNF_WELL_KNOWN"用于处理纯文本数据；"TNF_ABSOLUTE_URI"用于处理 URI 数据；"TNF_MIME_MEDIA"用于处理 MIME 媒介数据；等等。

NDEF 消息(NdefMessage)数据结构如图 7-4 所示。

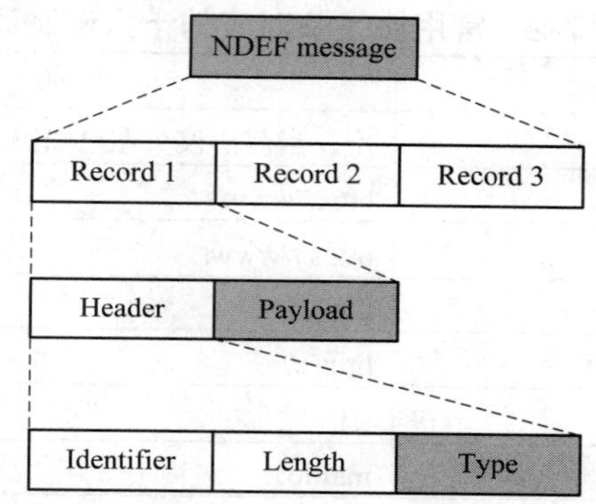

图 7-4　NDEF 消息(NdefMessage)数据结构图

每个记录(NdefRecord)分为头(Header)和有效负载(Payload)两部分，其中头部分用于定义记录的数据类型和长度等信息；有效负载部分用于存储要写入的数据。

NFC 数据类型被定义在 RTD 文档中， NFC Forum 组织主要定义了以下 RTD：

- ✧ NFC 纯文本 RTD，可携带 Unicode 字符串。文本记录包含在 NDEF 信息中，作为另一条记录的描述文本。
- ✧ NFC URI RTD，用于存储 URI 格式数据，例如网站地址、邮件、电话号码等，存储为经过优化的二进制形式。
- ✧ NFC 智能海报 RTD，用于将 URL、短信或电话号码编入 NFC Forum 标签，及如何在设备间传递这些信息。
- ✧ NFC 通用控制 RTD。
- ✧ NFC 签名 RTD。

其中，NDEF 纯文本数据以及 NDEF URI 数据相应的数据格式规范如下：

(1) NDEF 纯文本数据格式规范。NDEF 纯文本数据格式稍显复杂，整体(Payload)分为三部分：

- ✧ 第一部分：表示状态，占用第一个字节，每个字节有 8 位(0～7)。
  - ➢ 第 7 位(最后一位)表示编码格式，0 表示使用 UTF-8 编码，1 表示使用 UTF-16 编码。
  - ➢ 第 6 位为预留位，必须为 0。

➢ 第 0～5 位表示语言编码的长度(占用字节的个数)。

✧ 第二部分，语言编码，从第二个字节开始(1～n)，采用 ISO/IANA 语言编码，例如：fi、en-US、zh_CN 等。编码格式是 US-ASCII，长度由状态字节的后六位(0～5)指定。

✧ 第三部分，具体的文本数据(以上剩余部分)。

(2) NDEF URI 数据格式规范。NDEF URI 数据格式相对比较简洁，分为两个部分：

✧ 第一部分：占用一个字节(第 0 个字节)，用于存储已知的 URI 协议头部分，由于 NFC 标签容量有限，应当尽量减少不必要的空间浪费，因此协议头部分使用十六进制码表示。常用协议头对应的十六进制码如表 7-2 所示。

✧ 第二部分：用于存储 URI 除了协议头的剩余部分。

表 7-2 常用协议头对应的十六进制码

十六进制码	协 议 头
0x00	空，当前协议头在已知协议头列表中没有定义
0x01	http://www.
0x02	https://www.
0x03	http://
0x04	https://
0x05	tel:
0x06	mailto:
0x08	ftp://ftp.
0x09	ftps://
0x0D	ftp://
0x1D	file://

2. 非 NDEF 数据格式

表 7-1 中，除了 NDEF 技术的标签，其他均为非 NDEF 数据格式的标签，但需要注意的是：可以将 NdefFormatable 技术的标签转换为支持 NDEF 技术的标签，只需将内部存储的数据格式修改为 NDEF 数据格式即可；另外需要注意的是，一个标签可以支持多种 NFC 技术。

对于非 NDEF 数据格式的标签，本章主要讲解支持 MifareClassic 技术标签的读写操作。

通常，支持 MifareClassic 技术的标签，内存大小一般有 1 KB、2 KB 和 4 KB 三种，所拥有的分区(Sector)数分别为 16 个、32 个和 64 个，每个分区有 4 个块(Block)，每个块有 16 个 byte 数据。每个分区最后一个块叫作“Trailer”，主要用来存放读写该区的 Key，类似于密码。在读写每个分区数据之前，首先需要验证对应的 Key，只有通过验证，才能继续操作这个分区。Trailer 块分为 Key A(密码 A)、Access Conditions(存取控制)、Key B(密码 B)三部分，Key A 和 Key B 分别占 6 个 byte，默认的 Key 通常为 0xFF 或 0x00。其他块用于存储数据。

需要注意的是：块的下标从 0 开始，依次递增。例如：第一个分区四个块的下标为

0、1、2、3；第二个分区四个块的下标为 4、5、6、7，以此类推。也就是说，一个内存为 1 KB 的 MifareClassic 标签，有 64 个块，下标从 0 到 63。

7.3　Tag(标签)调度系统

当带有 NFC 功能的 Android 手机在开启 NFC 功能的状态下，扫描到 NFC 标签时，会自动读取该标签的信息，Tag 调度系统负责将读取到的数据格式和类型等信息封装为一个 Intent 对象，之后按照一定的规则进行调度分发，最终打开相应的 Activity 进行处理。如果有多个 Activity 同时匹配该标签，系统会列出程序列表供用户选择，这些 Activity 都需要在“AndroidManifest.xml”文件中注册对应的意图过滤器(Intent Filter)，NFC 标签过滤机制分为三个级别。这三个级别对应三个 Action 的定义：

- ✧ ACTION_NDEF_DISCOVERED：用于启动包含 NDEF 数据格式和已知类型的标签的 Activity。该 Intent 为最高级别，Android 会尽可能地使用这种类型的 Intent 来启动 Activity。
- ✧ ACTION_TECH_DISCOVERED：如果 ACTION_NDEF_DISCOVERED 类型的 Intent 没有被任何 Activity 注册处理，那么 Tag 调度系统会尝试使用该类型的 Intent 来启动对应的 Activity。如果被扫描的标签包含非 NDEF 数据，但是 NFC 技术是已知的，会直接启动注册该类型 Intent 的 Activity。
- ✧ ACTION_TAG_DISCOVERED：如果被扫描的标签不符合以上两种类型的 Intent 或者没有 Activity 注册，那么 Tag 调度系统会直接使用该类型的 Intent 启动 Activity。

总之，Tag 调度系统的基本工作流程如下：

(1) 解析扫描到的 NFC 标签数据，并封装为 Intent 对象。

(2) 使用封装好的 Intent 对象尝试启动对应的 Activity。

(3) 如果没有 Activity 来注册处理该 Intent，就会尝试使用下一个优先级的 Intent 来启动 Activity，直到有对应的 Activity 来处理这个 Intent。

(4) 如果最终还是没有 Activity 来注册处理该 Intent，那么就不做任何处理，最终结束调度。

通常建议尽量使用 NDEF 数据类型和 ACTION_NDEF_DISCOVERED 类型的 Intent 来进行程序的设计开发，因为该类型是上述三种 Intent 中最标准的，也是 Android API 所支持的格式。它允许开发者配置更加详尽的过滤机制来启动相应的 Activity，从而使程序具备更好的用户体验。

7.4　NFC 开发配置

在正式开发 NFC 应用程序之前，首先需要了解开发前的准备工作，主要体现在“AndroidManifest.xml”文件的配置方面。

1．添加 NFC 权限

在使用手机的 NFC 功能之前，需要在“AndroidManifest.xml”文件中声明 NFC 设备的使用权限，代码如下：

```
<uses-permission android:name="android.permission.NFC" />
```

2．设置程序所支持的最小 SDK 版本

为了更好地使用 NFC 功能，通常将程序支持的最小 SDK 版本设置为 API Leve 14，也就是 Android4.0 版本，代码如下：

```
<uses-sdk android:minSdkVersion="14"/>
```

3．在 Google Play 中声明支持 NFC 功能

使用“uses-feature”元素在 Google Play 中声明针对有 NFC 功能的设备中显示本程序，代码如下：

```
<uses-feature android:name="android.hardware.nfc" android:required="true" />
```

当然，如果本程序中的 NFC 功能不作为程序的关键功能，可以不对其进行设置。

4．配置意图过滤器

若要在手机扫描到 NFC 标签时，系统能够启动对应的应用程序，那么在程序的“AndroidManifest.xml”文件中需对指定的 Activity 的意图过滤器(Intent Filter)进行配置，可以针对一种或多种类型的 Intent 进行过滤。

(1) ACTION_NDEF_DISCOVERED。配置 ACTION_NDEF_DISCOVERED 级别的意图过滤器代码如下：

```
<intent-filter>
    <action android:name="android.nfc.action.NDEF_DISCOVERED"/>
    <category android:name="android.intent.category.DEFAULT"/>
    <data android:mimeType="text/plain" />
</intent-filter>
```

代码中<data>元素为声明只关心纯文本内容的 NDEF 格式数据，如果想关心所有内容的 NDEF 格式数据，此元素可省略。

如果只关心 URI 类型数据，例如一个网址，需要添加<data>元素，代码如下：

```
<data android:scheme="http"
    android:host="121ugrow.com"
    android:pathPrefix="/Index.aspx" />
```

当扫描到该标签时，将作为精准匹配来启动对应的 Activity，但如果找不到与之匹配的 Activity，系统会尝试寻找其他匹配的 Activity，如果最终没有任何 Activity 与之匹配，系统将调用浏览器打开<data>元素中定义的网址。

(2) ACTION_TECH_DISCOVERED。如果要过滤 ACTION_TECH_DISCOVERED 类型的 Intent，则需要重建一个 XML 资源文件，这个文件被放置于项目的“res/xml”文件夹中，文件名称可自定义，在该文件中的<tech-list>元素中定义所支持的 NFC 技术列表，如果定义的该“tech-list”集合是被扫描标签所支持的所有技术的一个子集，那么就认为

是匹配的。例如，被扫描的标签支持的技术包括“android.nfc.tech.NfcA”、“android.nfc.tech.Ndef”、“android.nfc.tech.MifareClassic”，那么想要匹配此类标签，定义“tech-list”集合的 XML 文件中至少包含其中一项技术，最多不能超过这三项技术，配置代码如下：

```
<resources xmlns:xliff="urn:oasis:names:tc:xliff:document:1.2">
    <tech-list>
        <tech>android.nfc.tech.NfcA</tech>
        <tech>android.nfc.tech.Ndef</tech>
        <tech>android.nfc.tech.MifareClassic</tech>
    </tech-list>
</resources>
```

另外，可以在此 XML 文件中定义多个“tech-list”集合，每个集合都是独立的。也就是说，每个集合都可以匹配一组标签技术，代码如下：

```
<resources xmlns:xliff="urn:oasis:names:tc:xliff:document:1.2">
    <tech-list>
        <tech>android.nfc.tech.NfcA</tech>
        <tech>android.nfc.tech.Ndef</tech>
        <tech>android.nfc.tech.MifareClassic</tech>
    </tech-list>
    <tech-list>
        <tech>android.nfc.tech. NfcA </tech>
        <tech>android.nfc.tech.NdefFormatable</tech>
    </tech-list>
</resources>
```

编辑完成 XML 文件后，最终需要在“AndroidManifest.xml”配置文件中对相应的 Activity 进行意图过滤器的配置。加入创建的用于匹配标签技术的 XML 文件名称为“nfc_tech_filter.xml”，意图过滤器的配置代码如下：

```
<activity ...>
    <intent-filter>
        <action android:name="android.nfc.action.TECH_DISCOVERED"/>
    </intent-filter>
    <meta-data android:name="android.nfc.action.TECH_DISCOVERED"
        android:resource="@xml/nfc_tech_filter" />
</activity>
```

(3) ACTION_TAG_DISCOVERED。过滤 ACTION_TAG_DISCOVERED 的配置代码如下：

```
<intent-filter>
    <action android:name="android.nfc.action.TAG_DISCOVERED"/>
</intent-filter>
```

7.5 NFC 标签数据操作

前面章节主要对 NFC 技术做了较为详细的介绍，包括 NFC 技术概述、工作模式、数据格式等。本小节将主要介绍 NFC 标签的数据操作——查看标签支持的 NFC 技术列表、读写 MifareClassic 标签数据和读写 NDEF 纯文本数据。

7.5.1 开发前的准备

Android API 中提供了开发 NFC 的相关类，较为重要的是 NfcAdapter 类，该类是程序与 NFC 硬件的桥梁，NfcAdapter 常用方法如表 7-3 所示。

表 7-3　NfcAdapter 常用方法

方　法　名	描　　述
disableForegroundDispatch(Activity activity)	禁用前台分发系统
disableForegroundNdefPush(Activity activity)	禁用点对点消息推送
enableForegroundDispatch(Activity activity, PendingIntent intent, IntentFilter[] filters, String[][] techLists)	启用前台分发系统
enableForegroundNdefPush(Activity activity, NdefMessage msg)	启用前台点对点消息推送
getDefaultAdapter(Context context)	获取系统默认 NFC 适配器
isEnabled()	检查 NFC 设备是否已启用

1. 读写 NFC 标签的主要步骤

读写 NFC 标签数据的主要步骤如下：

(1) 创建用于过滤标签技术的 XML 文件。

(2) 在 AndroidManifest.xml 文件中声明 NFC 权限，代码如下：

```
<uses-permission android:name="android.permission.NFC" />
```

(3) 将实现 NFC 功能的 Activity 设为单例模式启动。在<Activity>标签中，加入如下代码：

```
android:launchMode="singleTop"
```

(4) 如果必要，则在 AndroidManifest.xml 文件中进行过滤器的配置。

(5) 获取系统默认的 NfcAdapter 对象，代码如下：

```
NfcAdapternfcAdapter = NfcAdapter.getDefaultAdapter(this);
```

(6) 如果 NfcAdapter 对象为空，则说明当前手机不支持 NFC 功能，否则表示手机具有 NFC 功能。然后通过 NfcAdapter 对象查看手机是否已经开启了 NFC 功能，代码如下：

```
nfcAdapter.isEnabled();// true:已开启 NFC 功能，false:未开启 NFC 功能
```

(7) 如果要实现前台分发系统，则通常在 onResume()方法中开启，在 onPause()方法中禁用。

(8) 实现 onNewIntent()方法，当扫描到 NFC 标签后，以单例模式打开 Activity，执行此方法，在此方法中实现标签的相关操作。

(9) 最后对于具有不同标签技术和数据结构的标签进行不同的操作即可。

2．前台分发系统

简单地说，当程序(实现了 NFC 操作的 Activity)在前台运行时，可以通过前台分发系统将当前已启动的 Activity 拥有最高的优先级来过滤和处理 NFC 设备中获取的 Intent 对象，以便不受其他已注册相同过滤器的 Activity 的影响，甚至不会受到当前 Activity 在 AndroidManifest.xml 文件中已声明的意图过滤器(Intent Filter)的影响。也就是说，启用了前台分发系统的 Activity 的优先级大于 AndroidManifest.xml 文件中声明的意图过滤器。

通过 NfcAdapter.enableForegroundDispatch(Activity activity, PendingIntent intent, IntentFilter[] filters, String[][] techLists)方法可以启动前台分发系统，参数如下：

✧ Activity activity：要启动前台分发系统的 Activity。

✧ PendingIntent intent：分发时用到的 PendingIntent 对象。创建该对象的代码如下：

```
Intent intent = new Intent(this, getClass());
intent.addFlags(Intent.FLAG_ACTIVITY_SINGLE_TOP);
PendingIntent pendingIntent = PendingIntent.getActivity(this, 0, intent, 0);
```

✧ IntentFilter[] filters：意图过滤器集合，可为空。

✧ String[][] techLists：要过滤的标签技术列表，可为空。

启动前台分发的代码通常被添加到 onResume()方法中，而禁用前台分发的代码通常被添加到 onPause()方法中。

3．查看标签支持的 NFC 技术列表

在操作 NFC 标签数据之前，首先要了解当前被扫描的标签所支持的 NFC 技术，然后才能根据所支持的 NFC 技术来操作该标签。

下述示例用于实现：通过编程查看实验用到的 NFC 标签所支持的 NFC 技术列表。要求同时使用前台分发系统和“AndroidManifest.xml”文件注册的方式来实现此功能。当 Activity 关闭或运行于后台时，手机扫描到标签后，打开本程序以查看标签信息；当 Activity 运行于前台时，手机扫描标签后，直接显示标签信息，而不会受到其他程序的干扰。

(1) 创建项目“ch07_nfc_info”，在 res/xml 文件夹中创建 xml 资源文件“nfc_tech_filter.xml”。编写代码如下：

```
<resources xmlns:xliff="urn:oasis:names:tc:xliff:document:1.2">
<tech-list>
    <tech>android.nfc.tech.NfcA</tech>
</tech-list>
<tech-list>
    <tech>android.nfc.tech.MifareClassic</tech>
</tech-list>
<tech-list>
```

```
    <tech>android.nfc.tech.Ndef</tech>
</tech-list>
<tech-list>
    <tech>android.nfc.tech.NdefFormatable</tech>
  </tech-list>
</resources>
```

(2) 修改“activity_main.xml”布局文件，代码如下：

```
<LinearLayout xmlns:android="http://schemas.android.com/apk/res/android"
    xmlns:tools="http://schemas.android.com/tools"
    android:layout_width="match_parent"
    android:layout_height="match_parent"
    android:background="#ededed"
    android:orientation="vertical"
    android:padding="10dp" >

    <TextView
        android:id="@+id/act_main_info_tv"
        android:layout_width="wrap_content"
        android:layout_height="wrap_content"
        android:layout_marginTop="10dp"
        android:textSize="18sp" />
</LinearLayout>
```

在布局文件中添加一个 TextView 控件，用于显示信息。

(3) 在 AndroidManifest.xml 文件中进行注册，代码如下：

```
<?xml version="1.0" encoding="utf-8"?>
<manifest xmlns:android="http://schemas.android.com/apk/res/android"
    package="com.yg.ch07_nfc_info"
    android:versionCode="1"
    android:versionName="1.0" >

    <uses-sdk
        android:minSdkVersion="14"
        android:targetSdkVersion="14" />

<uses-permission android:name="android.permission.NFC" />

<application
        android:allowBackup="true"
        android:icon="@drawable/ic_launcher"
        android:label="@string/app_name"
```

```
        android:theme="@style/AppTheme" >
    <activity
        android:name=".MainActivity"
        android:label="@string/app_name"
        android:launchMode="singleTop" >
    <intent-filter>
    <action android:name="android.intent.action.MAIN" />

        <category android:name="android.intent.category.LAUNCHER" />
    </intent-filter>
    <intent-filter>
    <action android:name="android.nfc.action.NDEF_DISCOVERED" />
    </intent-filter>
    <intent-filter>
        <action android:name="android.nfc.action.TECH_DISCOVERED" />
    </intent-filter>
    <meta-data
        android:name="android.nfc.action.TECH_DISCOVERED"
            android:resource="@xml/nfc_tech_filter" />
    <intent-filter>
        <action android:name="android.nfc.action.TAG_DISCOVERED" />
    </intent-filter>
        </activity>
    </application>
</manifest>
```

(4) 修改 MainActivity.java 类，代码如下：

```
public class MainActivity extends Activity {
    private Tag tag = null;
    private NfcAdapter nfcAdapter = null;
    private TextView infoTv = null;
    // 启用前台分配的 PendingIntent 对象
    private PendingIntent pendingIntent = null;

    @Override
    protected void onCreate(Bundle savedInstanceState) {
        super.onCreate(savedInstanceState);
        setContentView(R.layout.activity_main);

        infoTv = (TextView) findViewById(R.id.act_main_info_tv);
```

```
        // 获取系统默认的 NFC 设备
        nfcAdapter = NfcAdapter.getDefaultAdapter(this);
        // 如果为空，则说明手机不支持 NFC
        if (nfcAdapter == null) {
                infoTv.setText("该手机不支持 NFC 功能");
                return;
        }

        Intent intent = new Intent(this, getClass());
        intent.addFlags(Intent.FLAG_ACTIVITY_SINGLE_TOP);
        pendingIntent = PendingIntent.getActivity(this, 0, intent, 0);
}

@Override
protected void onResume() {
        super.onResume();
        if (nfcAdapter == null) {
                return;
        }
        // 检查是否开启 NFC 功能
        if (!nfcAdapter.isEnabled()) {
                infoTv.setText("请在设置中开启 NFC 功能");
                return;
        } else {
                infoTv.setText("NFC 功能已开启，等待扫描中...\n");
                // 调用 onNewIntent()并传入当前 Intent 实例
                onNewIntent(getIntent());
                // 启用前台分发
                nfcAdapter.enableForegroundDispatch(this,
                        pendingIntent, null, null);
        }
}

@Override
protected void onNewIntent(Intent intent) {
        setIntent(intent);
        String action = intent.getAction();
        String[] techList = null;
        if (NfcAdapter.ACTION_NDEF_DISCOVERED.equals(action)
                        || NfcAdapter.ACTION_TECH_DISCOVERED.equals(action)
```

```
                || NfcAdapter.ACTION_TAG_DISCOVERED.equals(action)) {
            // Action
            infoTv.setText("Action：\n" + action + "\n");
            // 标签 ID
            byte[] tagId = intent.getByteArrayExtra(NfcAdapter.EXTRA_ID);
            infoTv.append("标签 ID：" + bytesToHexString(tagId) + "\n");

            // 获取 NFC 标签对象
            tag = intent.getParcelableExtra(NfcAdapter.EXTRA_TAG);
            // 获取支持的技术列表
            techList = tag.getTechList();
            infoTv.append("该 NFC 标签所支持的技术有:\n");
            for (String tech : techList) {
                infoTv.append(tech + "\n");
            }
        }
}

@Override
protected void onPause() {
        super.onPause();
        if (nfcAdapter != null) {
            if (nfcAdapter.isEnabled()) {
                // 禁用前台分发
                nfcAdapter.disableForegroundDispatch(this);
            }
        }
}

/**
 * 将 byte 数组转为字符串
 *
 * @param bytes
 * @return
 */
private String bytesToHexString(byte[] bytes) {

        int i, j, t;
        String[] hex = { "0", "1", "2", "3", "4", "5", "6", "7",
            "8", "9", "A","B", "C", "D", "E", "F" };
```

```
            String r = "";

            for (j = 0; j < bytes.length; ++j) {
                    t = bytes[j] & 0xff;
                    i = (t >> 4) & 0x0f;
                    r += hex[i];
                    i = t & 0x0f;
                    r += hex[i];
            }
            return r;
    }
}
```

在上述代码中，MainActivity 以单例模式运行，当启动程序时，首先判断当前手机是否具有 NFC 功能，然后判断是否开启了 NFC 功能。核心代码位于 onNewIntent()回调方法中。

(5) 运行程序后，使当前程序位于前台，扫描标签后，将直接显示标签信息，如图 7-5 所示；将当前程序关闭或者位于后台，再次扫描标签，如果手机中有多个 NFC 读取软件，就会出现如图 7-6 所示的程序列表，打开名称为“ch07_nfc_info”的程序，然后显示标签信息。

图 7-5　扫描结果

图 7-6　扫描后显示列表

7.5.2　读写 MifareClassic 标签数据

写入 MifareClassic 格式数据时，每个块(Block)只能写入 16 个字节(byte)，如果不够 16 个字节，将剩余的高位补 0 即可。当然，如果想一次性写入超过 16 个字节的数据，也可以通过编程来实现。方案是将写入的数据进行分段，之后依次写入多个块中，必须要注意的是：每个扇区(Sector)最后一个块为校验块，除非必要，否则不建议修改。

读写 MifareClassic 标签数据的基本步骤如下：

(1) 当在 onNewIntent()方法中捕获到了相关 Action 之后，通过 Intent 对象的 getParcelableExtra()方法获取 Tag 对象，代码如下：

```
Tag tag = intent.getParcelableExtra(NfcAdapter.EXTRA_TAG);
```

(2) 通过 Tag 对象获取 MifareClassic 对象，代码如下：

```
MifareClassic mcNfc = MifareClassic.get(tag);
```

(3) 与标签建立连接，代码如下：

```
mcNfc.connect();
```

(4) 之后即可对标签进行读写操作；

(5) 操作完毕后，关闭与标签的连接，代码如下：

```
mcNfc.close();
```

下述示例用于实现：通过编程实现 MifareClassic 格式标签数据的读写功能。要求在“查看标签支持的 NFC 技术列表”示例的基础上实现：显示标签的基本数据，包括扇区数量、总块数、标签容量等；对指定的扇区和块写入用户输入的文本；查看所有扇区的数据；清空所有数据。

(6) 创建项目“ch07_nfc_mifarec_io”，在 res/xml 文件夹中创建 xml 资源文件“nfc_tech_filter.xml”，代码同“查看标签支持的 NFC 技术列表”示例中代码，此处不再赘述。

(7) 修改“activity_main.xml”布局文件，代码如下：

```
<LinearLayout xmlns:android="http://schemas.android.com/apk/res/android"
    xmlns:tools="http://schemas.android.com/tools"
    android:layout_width="match_parent"
    android:layout_height="match_parent"
    android:background="#ededed"
    android:orientation="vertical"
    android:padding="10dp" >

    <EditText
        android:id="@+id/act_main_sector_et"
        android:layout_width="fill_parent"
        android:layout_height="wrap_content"
        android:hint="待写入扇区索引"
        android:inputType="number"
        android:singleLine="true" />

    <EditText
        android:id="@+id/act_main_block_et"
        android:layout_width="fill_parent"
        android:layout_height="wrap_content"
        android:hint="待写入块索引(0-2)"
        android:inputType="number"
```

```
        android:singleLine="true" />

    <EditText
        android:id="@+id/act_main_content_et"
        android:layout_width="fill_parent"
        android:layout_height="wrap_content"
        android:hint="待写入内容" />

    <LinearLayout
        android:layout_width="match_parent"
        android:layout_height="wrap_content"
        android:orientation="horizontal" >

        <Button
            android:id="@+id/act_main_write_btn"
            android:layout_width="fill_parent"
            android:layout_height="wrap_content"
            android:layout_weight="1"
            android:text="写入" />

        <Button
            android:id="@+id/act_main_read_btn"
            android:layout_width="fill_parent"
            android:layout_height="wrap_content"
            android:layout_weight="1"
            android:text="读取" />

        <Button
            android:id="@+id/act_main_clear_btn"
            android:layout_width="fill_parent"
            android:layout_height="wrap_content"
            android:layout_weight="1"
            android:text="清空" />
    </LinearLayout>

    <ScrollView
        android:layout_width="fill_parent"
        android:layout_height="fill_parent" >

        <LinearLayout
```

```
                android:layout_width="fill_parent"
                android:layout_height="wrap_content"
                android:orientation="vertical" >

                <TextView
                        android:id="@+id/act_main_info_tv"
                        android:layout_width="wrap_content"
                        android:layout_height="wrap_content"
                        android:layout_marginTop="10dp"
                        android:textSize="18sp" />

                <TextView
                        android:id="@+id/act_main_content_tv"
                        android:layout_width="wrap_content"
                        android:layout_height="wrap_content"
                        android:layout_marginTop="10dp"
                        android:textSize="18sp" />
        </LinearLayout>
    </ScrollView>

</LinearLayout>
```

(8) 在 AndroidManifest.xml 文件中进行注册，代码如下：

```
<?xml version="1.0" encoding="utf-8"?>
<manifest xmlns:android="http://schemas.android.com/apk/res/android"
    package="com.yg.ch07_nfc_mifarec_io"
    android:versionCode="1"
    android:versionName="1.0" >

    <uses-sdk
        android:minSdkVersion="14"
        android:targetSdkVersion="14" />

    <uses-permission android:name="android.permission.NFC" />

    <application
        android:allowBackup="true"
        android:icon="@drawable/ic_launcher"
        android:label="@string/app_name"
        android:theme="@style/AppTheme" >
        <activity
```

```
                android:name=".MainActivity"
                android:label="@string/app_name"
                android:launchMode="singleTop" >
                <intent-filter>
                    <action android:name="android.intent.action.MAIN" />

                    <category android:name="android.intent.category.LAUNCHER" />
                </intent-filter>
                <intent-filter>
                    <action android:name="android.nfc.action.NDEF_DISCOVERED" />
                </intent-filter>
                <intent-filter>
                    <action android:name="android.nfc.action.TECH_DISCOVERED" />
                </intent-filter>

                <meta-data
                    android:name="android.nfc.action.TECH_DISCOVERED"
                    android:resource="@xml/nfc_tech_filter" />

                <intent-filter>
                    <action android:name="android.nfc.action.TAG_DISCOVERED" />
                </intent-filter>
            </activity>
        </application>

</manifest>
```

(9) “MainActivity.java”类中的代码将分步骤来实现，首先实现基本代码，包括布局文件中控件的引用、必要的生命周期方法等，代码如下：

```
public class CopyOfMainActivity extends Activity {


    private TextView infoTv;
    /** 待写入扇区号 */
    private EditText sectorEt;
    /** 待写入块号 */
    private EditText blockEt;
    /** 待写入内容 */
    private EditText contentEt;
    /** 显示读取数据结果 */
    private TextView contentTv;
    private Button writeBtn;
```

```
    private Button readBtn;
    private Button clearBtn;

    private NfcAdapter nfcAdapter;
    private PendingIntent pendingIntent;
    private Tag tag;
    private MifareClassic mcNfc;

    @Override
    protected void onCreate(Bundle savedInstanceState) {
        super.onCreate(savedInstanceState);
        setContentView(R.layout.activity_main);

        infoTv = (TextView) findViewById(R.id.act_main_info_tv);
        sectorEt = (EditText) findViewById(R.id.act_main_sector_et);
        blockEt = (EditText) findViewById(R.id.act_main_block_et);
        contentEt = (EditText) findViewById(R.id.act_main_content_et);
        contentTv = (TextView) findViewById(R.id.act_main_content_tv);

        writeBtn = (Button) findViewById(R.id.act_main_write_btn);
        readBtn = (Button) findViewById(R.id.act_main_read_btn);
        clearBtn = (Button) findViewById(R.id.act_main_clear_btn);
        writeBtn.setOnClickListener(onBtnClickListener);
        readBtn.setOnClickListener(onBtnClickListener);
        clearBtn.setOnClickListener(onBtnClickListener);

        // 获取系统默认的 NFC 设备
        nfcAdapter = NfcAdapter.getDefaultAdapter(this);
        // 如果为空，则说明手机不支持 NFC
        if (nfcAdapter == null) {
            contentTv.setText("该手机不支持 NFC 功能");
            return;
        }

        Intent intent = new Intent(this, getClass());
        intent.addFlags(Intent.FLAG_ACTIVITY_SINGLE_TOP);
        pendingIntent = PendingIntent.getActivity(this, 0, intent, 0);
    }

    @Override
```

```
    protected void onResume() {
        super.onResume();
        //代码略，内容同 ch07_nfc_info
    }

    @Override
    protected void onPause() {
        super.onPause();
        //代码略，内容同 ch07_nfc_info
    }

    @Override
    protected void onNewIntent(Intent intent) {
        //代码略，内容同 ch07_nfc_info

        // 在 ch07_nfc_info 项目基础上加入的新代码
        // 用于将扇区数显示到 EditText
        if (techList != null) {
            showSectors(techList);
        }
    }
}
```

上述代码，主要实现了基本布局的引用和生命周期方法代码，其中，三个按钮的点击事件还未实现。在“ch07_nfc_info”项目的基础上，onNewIntent()方法增加了一段代码，showSectors()方法用于将扇区数显示到 EditText；另外，onNewIntent()方法中还需要实现 bytesToHexString()方法，因为解析获取的标签 ID 为 byte 数组，此方法将 byte 数组转换为 String 并返回。bytesToHexString()方法同“ch07_nfc_info”项目，此处不再赘述。

(10) 实现 showSectors()方法，代码如下：

```
/**
 * 将扇区数显示到 EditText
 *
 * @param techList
 */
private void showSectors(String[] techList) {
    // 转为 List 列表对象
    List<String> list = Arrays.asList(techList);
    if (list.contains("android.nfc.tech.MifareClassic")) {
        mcNfc = MifareClassic.get(tag);
        try {
            mcNfc.connect();
```

```
                int sectorCnt = mcNfc.getSectorCount();
                sectorEt.setHint("待写入扇区索引(共 " + sectorCnt + " 个扇区)");
            } catch (Exception e) {
                contentTv.setText("连接标签失败" + e.getMessage());
                e.printStackTrace();
            } finally {
                try {
                    mcNfc.close();
                } catch (IOException e) {
                    e.printStackTrace();
                }
            }
        } else {
            contentTv.append("该标签不支持 MifareClassic 技术");
        }
}
```

上述代码主要功能是将读取到的扇区数显示到 EditText 控件上，以便提示用户可写入的最大扇区号，通过“list.contains("android.nfc.tech.MifareClassic")”判断标签是否支持 MifareClassic 技术，只有支持这项技术的情况下，才可以继续读取信息。

(11) 实现按钮点击事件，代码如下：

```
private OnClickListener onBtnClickListener = new OnClickListener() {

        @Override
        public void onClick(View v) {
            if (v == writeBtn) {
                // 获取用户输入的扇区、块和内容
                String sectorStr = sectorEt.getText().toString();
                String blockStr = blockEt.getText().toString();
                String contentStr = contentEt.getText().toString();

                if (sectorStr.isEmpty() || blockStr.isEmpty()) {
                    Toast.makeText(getApplicationContext(),
                            "请填写扇区以及块索引",Toast.LENGTH_SHORT).show();
                    return;
                }
                // 将扇区和块索引转为 int 类型
                int sector = Integer.parseInt(sectorStr);
                int block = Integer.parseInt(blockStr);
                // 开始写入标签
                writeNFC(sector, block, contentStr);
```

```
            } else if (v == readBtn) {
                    // 读取标签数据
                    readNFC();
            } else if (v == clearBtn) {
                    // 清空标签数据
                    clearNFC();
            }
    }
};
```

在上述代码中，以全局成员的方式声明了按钮的 OnClickListener 事件，该事件实现了写入、读取和清空三个按钮的监听。这三个按钮具体功能的实现分别需要调用三个方法：writeNFC()、readNFC()、clearNFC()。其中，在“写入”按钮中，需要验证用户输入的内容，验证成功后，才能继续进行写入数据操作。

(12) 实现数据的写入，添加 writeNFC()方法，代码如下：

```
/**
 * 将文本数据写入标签
 *
 * @param sectorIdx   扇区下标
 *
 * @param blockIdx   块的下标(0-3)
 *
 * @param content   文本数据
 *
 */
protected void writeNFC(int sectorIdx, int blockIdx, String content) {
        try {
                mcNfc.connect();
                boolean auth = false;
                // 对扇区进行验证
                auth = mcNfc.authenticateSectorWithKeyA(sectorIdx,
                                MifareClassic.KEY_NFC_FORUM);
                if (!auth) {
                        contentTv.setText("验证扇区失败");
                        return;
                }
                // 计算需要写入的 Block 位置
                blockIdx = calcBlockId(sectorIdx, blockIdx);
                // 将要写入的内容转为 byte 数组，数据不足 16byte，高位剩余补 0
                byte[] data = calcDataTo16Byte(content.getBytes());
                if (data.length > 16) {
```

```
                    contentTv.setText("写入数据过多");
                    return;
                }
                mcNfc.writeBlock(blockIdx, data);
                contentTv.setText("写入数据成功");

        } catch (Exception e) {
                contentTv.setText("写入标签失败：" + e.getMessage());
                e.printStackTrace();
        } finally {
                try {
                        mcNfc.close();
                } catch (IOException e) {
                        e.printStackTrace();
                }
        }
}
```

上述代码主要实现了将用户输入的文本信息写入指定扇区内的指定块中。在写入数据之前，需要通过 MifareClassic.connect()方法连接标签，连接成功后，验证要操作的扇区，之后才能对指定的块进行写入数据的操作。需要注意的是：

- ✧ 用户输入的块索引是从 0～2 之间的数字(是相对于当前扇区的)，因此程序中需要根据扇区的索引和块的索引计算出要写入块的绝对索引位置，调用 calcBlockId()方法进行计算；
- ✧ 每个块只能写入 16 字节数据，如果不足 16 个字节，则需要将高位进行“补 0”操作，通过调用 calcDataTo16Byte()方法进行计算。

(13) 实现 calcBlockId()和 calcDataTo16Byte()方法，代码如下：

```
/**
 * 计算需要写入的 Block 位置
 *
 * @param sectorIdx
 * @param blockIdx
 * @return
 */
private int calcBlockId(int sectorIdx, int blockIdx) {
        if (sectorIdx == 0) {
                return blockIdx;
        }
        return sectorIdx * 4 + blockIdx;
}
```

```
/**
 * 如果数据不足 16 位，则将高位剩余补 0
 *
 * @param contentBytes
 * @return
 */
private byte[] calcDataTo16Byte(byte[] contentBytes) {
        if (contentBytes.length >= 16) {
                return contentBytes;
        }
        byte[] data = new byte[16];
        int contentLength = contentBytes.length;
        System.arraycopy(contentBytes, 0, data, 16 - contentLength,
                        contentLength);
        return data;

}
```

至此，向标签中写入数据的工作已经完成，运行程序进行测试，但是读取数据的功能还没有完成，因此无法查看是否真正写入了数据。

(14) 实现读取数据的功能(readNFC()方法)，代码如下：

```
/**
 * 读取标签数据
 */
protected void readNFC() {
        String str = "";
        if (mcNfc == null) {
                contentTv.setText("NFC 标签不支持 MifareClassic 技术！");
                return;
        }
        try {
                mcNfc.connect();
                int sectorCnt = mcNfc.getSectorCount();
                int blockCnt = mcNfc.getBlockCount();
                int size = mcNfc.getSize();
                int timout = mcNfc.getTimeout();
                int maxCmdLen = mcNfc.getMaxTransceiveLength();

                str += "扇区数量:" + sectorCnt + "\n";
                str += "总块数:" + blockCnt + "\n";
                str += "标签容量:" + size + "\n";
```

```
            str += "超时时间:" + timout + "\n";
            str += "最大指令长度:" + maxCmdLen + "\n";
            contentTv.setText(str);

            // 读取标签数据
            StringBuffer sb = new StringBuffer();
            // 遍历 Sector(扇区)
            for (int i = 0; i < sectorCnt; i++) {
                sb.append("第 " + i + " 块扇区:\n");
                boolean auth = false;
                // 对每块扇区进行验证
                auth = mcNfc.authenticateSectorWithKeyA(i,
                            MifareClassic.KEY_NFC_FORUM);
                // 验证通过，遍历 Block(块)，每个扇区有 4 个 Block
                if (auth) {
                    // 扇区下标*4 就是当前需要遍历的第一个 Block
                    int k = i * 4;
                    // 每个 Block 有 16 个 byte，以 byte[]形式保存
                    // 每个扇区的最后一个 Block 用作校验，不用作保存数据
                    for (int j = k, z = 0; j < k + 4; j++, z++) {
                        byte[] item = mcNfc.readBlock(j);
                        sb.append("\t 第 " + j + " 个块: ");
                        if (z == 3) {
                            sb.append("- 校验块 -");
                        }
                        // sb.append(bytesToHexString(item) + "\n");
                        sb.append(new String(item) + "\n");
                    }
                } else {
                    sb.append("\t 认证失败\n");
                }
            }
            contentTv.append(sb.toString());
        } catch (Exception e) {
            contentTv.setText("读取标签失败：" + e.getMessage());
            e.printStackTrace();
        } finally {
            try {
                mcNfc.close();
            } catch (IOException e) {
```

```
                    e.printStackTrace();
                }
            }
}
```

上述代码实现了 NFC 标签的读取数据功能。与写入数据类似，必须先连接标签，只有连接成功后才能继续操作。连接成功后，获取标签的信息，例如：扇区数量、总块数、标签容量等。核心代码在读取数据部分，首先遍历每个分区，对每个扇区进行验证，验证通过后，遍历当前扇区的所有块，将最后一个块显示为“校验块”文本，用于表示校验块。

(15) 运行程序后，读取之前写入的数据，以验证是否成功。写入数据如图 7-7 所示，读取数据如图 7-8 所示。

图 7-7　写入数据

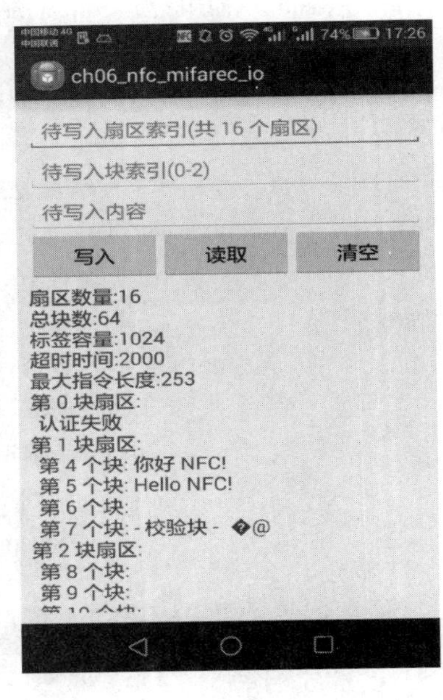

图 7-8　读取数据

(16) 实现清空数据功能。清空数据就是遍历每一个块，并写入长度为 16 的 byte 空数组，可以将 writeNFC()方法稍作修改，加以重用。本示例为了简洁易懂，不做重用处理。clearNFC()方法的代码如下：

```
/**
 * 清空所有数据
 */
protected void clearNFC() {
        if (mcNfc != null) {
                try {
                        mcNfc.connect();
                        int sectorCnt = mcNfc.getSectorCount();
```

```
                byte[] emptyByte = new byte[16];
                StringBuffer sb = new StringBuffer();
                // 遍历 Sector(扇区)
                for (int i = 0; i < sectorCnt; i++) {
                        sb.append("第 " + i + " 块扇区:\n");
                        boolean auth = false;
                        // 对每块扇区进行验证
                        auth = mcNfc.authenticateSectorWithKeyA(i,
                                        MifareClassic.KEY_NFC_FORUM);// 验证密码
                        if (auth) {
                                int k = i * 4;
                                for (int j = k, z = 0; j < k + 4; j++, z++) {
                                        sb.append("\t 第 " + j + " 个块: ");
                                        // 每个扇区的第三个 Bolck 用作校验，谨慎修改
                                        //此处不做修改
                                        if (z == 3) {
                                                sb.append("- 校验块 不做改动 -");
                                                break;
                                        }
                                        mcNfc.writeBlock(j, emptyByte);
                                        sb.append("已清空\n");
                                }
                        } else {
                                sb.append("\t 认证失败\n");
                        }
                        sb.append("\n");
                }
                contentTv.setText(sb.toString());
        } catch (Exception e) {
                contentTv.setText("清空数据失败:" + e.getMessage());
                e.printStackTrace();
        } finally {
                try {
                        mcNfc.close();
                } catch (IOException e) {
                        e.printStackTrace();
                }
        }
    }
}
```

上述代码实现了清空标签数据的功能。清空数据也就是写入空数据，不再赘述。运行程序后，点击“清空”按钮，再次读取数据，验证是否成功清空数据。清空数据后的结果如图 7-9 所示。至此，读写 MifareClassic 标签数据功能已完成。

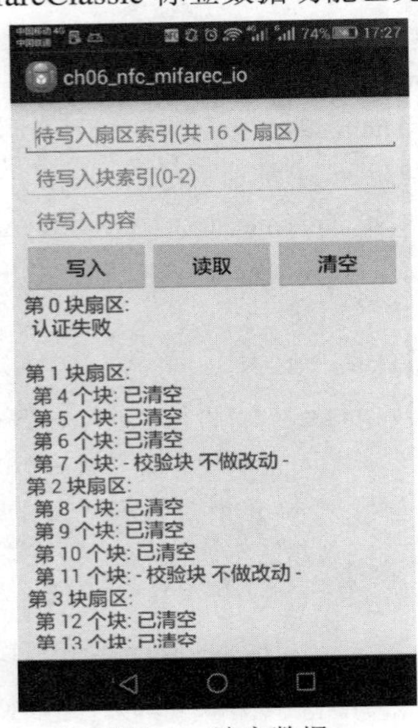

图 7-9　清空数据

7.5.3　读写 NDEF 纯文本数据

相对于读写 MifareClassic 数据，读写 NDEF 纯文本数据更加简洁，没有 16 位字节的限制、没有扇区和块的概念，数据被封装到 NdefRecord 对象中进行写入操作。

下述示例用于实现：通过编程实现 NDEF 格式标签纯文本数据的读写功能。要求在“查看标签支持的 NFC 技术列表”示例的基础上实现：如果标签技术支持的是 NdefFormatable 技术，则将其转换为支持 NDEF 技术，以支持 NDEF 数据格式；能够将用户输入的数据写入到标签；能够读取标签中保存的数据。

(1) 创建项目“ch07_nfc_ndef_io”，在“res/xml”文件夹中创建 XML 资源文件“nfc_tech_filter.xml”，代码同“查看标签支持的 NFC 技术列表”示例中相应代码，此处不再赘述。

(2) 修改“activity_main.xml”布局文件，代码如下：

```
<LinearLayout xmlns:android="http://schemas.android.com/apk/res/android"
    xmlns:tools="http://schemas.android.com/tools"
    android:layout_width="match_parent"
    android:layout_height="match_parent"
    android:background="#ededed"
    android:orientation="vertical"
    android:padding="10dp" >
```

```
    <Button
        android:id="@+id/act_main_convert_btn"
        android:layout_width="fill_parent"
        android:layout_height="wrap_content"
        android:text="转换为标准 NDEF 格式数据" />

    <EditText
        android:id="@+id/act_main_content_et"
        android:layout_width="fill_parent"
        android:layout_height="wrap_content"
        android:layout_marginTop="10dp"
        android:hint="要写入的数据" />

<LinearLayout
        android:layout_width="fill_parent"
        android:layout_height="wrap_content"
        android:orientation="horizontal" >

<Button
        android:id="@+id/act_main_writendef_btn"
        android:layout_width="wrap_content"
        android:layout_height="wrap_content"
        android:layout_weight="1"
        android:text="写 NDEF 数据" />

    <Button
        android:id="@+id/act_main_readndef_btn"
        android:layout_width="wrap_content"
        android:layout_height="wrap_content"
        android:layout_weight="1"
        android:text="读 NDEF 数据" />
    </LinearLayout>

    <TextView
        android:id="@+id/act_main_info_tv"
        android:layout_width="wrap_content"
        android:layout_height="wrap_content"
        android:layout_marginTop="10dp"
        android:textSize="18sp" />
```

```
    <TextView
        android:id="@+id/act_main_content_tv"
        android:layout_width="wrap_content"
        android:layout_height="wrap_content"
        android:layout_marginTop="10dp"
        android:textSize="18sp" />
</LinearLayout>
```

(3) 在 AndroidManifest.xml 文件中进行注册，代码与“ch07_nfc_mifarec_io”项目中相应代码类似，此处不再赘述。

(4) MainActivity.java 类中的代码将分步骤来实现，基本代码包括布局文件中控件的引用、必要的生命周期方法等。代码如下：

```
public class MainActivity extends Activity {
    private Tag tag = null;
    private NfcAdapter nfcAdapter = null;

    private TextView infoTv = null;
    private TextView contentTv = null;

    private EditText contentEt = null;

    private Button convertBtn = null;
    private Button writeNdefBtn = null;
    private Button readNdefBtn = null;

    // 启用前台分配的 PendingIntent 对象
    private PendingIntent pendingIntent = null;

    @Override
    protected void onCreate(Bundle savedInstanceState) {
        super.onCreate(savedInstanceState);
        setContentView(R.layout.activity_main);

        infoTv = (TextView) findViewById(R.id.act_main_info_tv);
        contentTv = (TextView) findViewById(R.id.act_main_content_tv);
        contentEt = (EditText) findViewById(R.id.act_main_content_et);
        convertBtn = (Button) findViewById(R.id.act_main_convert_btn);
        writeNdefBtn =
                (Button) findViewById(R.id.act_main_writendef_btn);
        readNdefBtn = (Button) findViewById(R.id.act_main_readndef_btn);
```

```
            convertBtn.setOnClickListener(onBtnClickListener);
            writeNdefBtn.setOnClickListener(onBtnClickListener);
            readNdefBtn.setOnClickListener(onBtnClickListener);

            // 获取系统默认的 NFC 设备
            nfcAdapter = NfcAdapter.getDefaultAdapter(this);
            // 如果为空，说明手机不支持 NFC
            if (nfcAdapter == null) {
                    contentTv.setText("该手机不支持 NFC 功能");
                    return;
            }

            Intent intent = new Intent(this, getClass());
            intent.addFlags(Intent.FLAG_ACTIVITY_SINGLE_TOP);
            pendingIntent = PendingIntent.getActivity(this, 0, intent, 0);
    }
    @Override
    protected void onResume() {
            super.onResume();
            //代码略，内容同 ch07_nfc_info
    }
    @Override
    protected void onPause() {
            super.onPause();
            //代码略，内容同 ch07_nfc_info
    }
    @Override
    protected void onNewIntent(Intent intent) {
            // 代码略，内容同 ch07_nfc_info
    }
}
```

上述代码，主要实现了基本的布局的引用和生命周期方法，其中，三个按钮的点击事件还未实现。在 onNewIntent()方法中还需要实现 bytesToHexString()方法，之前项目中已有介绍，此处不再赘述。

(5) 实现按钮点击事件，代码如下：

```
private OnClickListener onBtnClickListener = new OnClickListener() {

    @Override
    public void onClick(View v) {
            if (v == convertBtn) {
```

```
                convertToNdef();
            } else if (v == writeNdefBtn) {
                String content = contentEt.getText().toString();
                writeNdef(content);
            } else if (v == readNdefBtn) {
                readNdef();
            }
        }

};
```

上述代码实现方式和“读写 MifareClassic 标签数据”示例中方式基本类似，此处也不赘述。该事件实现了对三个按钮的监听：“convertBtn”按钮实现将 NdefFormatable 技术的标签转换为 NDEF 数据格式；“writeNdefBtn”按钮用于将用户输入的文本写入标签；“readNdefBtn”按钮用于读取标签数据。分别调用 convertToNdef()、writeNdef()和 readNdef()方法实现三个按钮的功能。

(6) 实现 convertToNdef()方法，将 NdefFormatable 技术的标签转换为 NDEF 数据格式，代码如下：

```
/**
 * 将 NdefFormatable 格式转换为 NDEF 格式
 */
private void convertToNdef() {
    // 获取 NdefFormatable 对象
    NdefFormatable ndefFormat = NdefFormatable.get(tag);

    // 只有 NdefFormatable 对象不为空时，才能尝试写入 NDEF 格式数据
    if (ndefFormat == null) {
        contentTv.setText("NFC 标签不支持 NdefFormatable 技术！");
        return;
    }

    try {
        // 连接标签
        ndefFormat.connect();
        // 因为目的是转换为 NDEF 格式，因此这里创建空数据即可
        NdefRecord record = createTextRecord("");
        NdefRecord[] records = new NdefRecord[] { record };
        // 通过 records 创建 NdefMessage 对象
        NdefMessage ndefMessage = new NdefMessage(records);

        // 向标签写入数据
```

```
            ndefFormat.format(ndefMessage);
            contentTv.setText("已成功写入数据！");

        } catch (Exception e) {
            contentTv.setText("写入数据失败！");
            e.printStackTrace();
        } finally {
            try {
                // 关闭与标签的连接
                ndefFormat.close();
            } catch (IOException e) {
                e.printStackTrace();
            }
        }

}
```

在上述代码中，首先将之前获取的 Tag 对象转换为 NdefFormatable 对象，如果失败，则说明该标签不支持 NdefFormatable 技术。因为数据被封装到 NdefRecord 中，所以需要通过 createTextRecord()将要写入的文本数据封装为 NdefRecord 对象并返回。之后通过 NdefRecord 对象创建 NdefMessage 对象。最后将 NdefMessage 对象写入标签中。

(7) 实现 createTextRecord()方法，将要写入的文本数据封装为 NdefRecord 对象。此方法为核心代码，代码如下：

```
/**
 * 将要写入的数据封装为一个 NdefRecord 对象
 *
 * @param content
 * @return
 */
public NdefRecord createTextRecord(String content) {
    // 指定语言编码为中文，转为 byte 数组
    byte[] langBytes = Locale.CHINA.getLanguage().getBytes(
                Charset.forName("US-ASCII"));
    // 设置编码方式为 UTF-8，并将其转换为 byte 数组
    Charset utfEncoding = Charset.forName("UTF-8");
    // 将要写入的文本数据转换为 byte 数组
    byte[] contentBytes = content.getBytes(utfEncoding);
    // 0：utf-8. 1：utf-16
    int utf = 0;

    // 语言编码的长度
```

```
        int langLength = langBytes.length;
        // 将要写入的文本数据长度
        int contentLength = contentBytes.length;

        // 第一部分的状态字符，转为 byte
        char statusChar = (char) (utf + langLength);
        byte statusByte = (byte) statusChar;

        // 创建存储 Payload 的字节数组
        // 1 表示第一位保存状态字节，第二部分语言编码，第三部分文本数据的长度
        byte[] data = new byte[1 + langLength + contentLength];
        // 将 Payload 的第一位存入状态字节
        data[0] = statusByte;
        // 将语言编码部分的 byte[]复制到 Payload 字节数组中，
        //位置是从 1~语言编码的长度位置
        System.arraycopy(langBytes, 0, data, 1, langLength);
        // 将要写入的文本数据复制到 Payload 字节数组中，位置是从语言编码之后的位置开始写入
        System.arraycopy(contentBytes, 0, data, 1 + langLength,
                contentLength);

        // 使用创建好的 Payload 数据创建 NdefRecord 对象
        NdefRecord record = new NdefRecord(NdefRecord.TNF_WELL_KNOWN,
                        NdefRecord.RTD_TEXT, new byte[0], data);
        return record;
}
```

上述代码作为程序的核心代码，主要将文本数据根据 NDEF 数据格式封装成标准的 NdefRecord 对象，代码中有详细的注释，在此不再赘述。

至此，已完成了将支持 NdefFormatable 技术的标签转换为支持 NDEF 技术的标签，运行程序并扫描标签，如果支持的技术列表中没有“NDEF”，而有“NdefFormatable”，那么就可以通过点击“转换为标准 NDEF 格式数据”按钮，将其进行转换。需要注意的是：最终目的是转换格式，并不需要关心写入的文本内容。因此，这里默认写入空白数据。将 NdefFormatable 转换为 NDEF 技术的结果如图 7-10 所示，再次扫描标签，会发现“技术列表”中已经存在支持 NDEF 的技术。

图 7-10　转换后的结果

(8) 实现 writeNdef()方法，此方法可将 NDEF 纯文本数据写入到标签，代码如下：

```
/**
 * 将文本数据写入到标签
 *
 * @param content
 */
private void writeNdef(String content) {
        // 创建 record 并封装文本数据
        NdefRecord record = createTextRecord(content);
        NdefRecord[] records = new NdefRecord[] { record };
        // 通过 records 创建 NdefMessage 对象
        NdefMessage ndefMessage = new NdefMessage(records);
        // 获取 NdefMessage 对象字节长度
        int ndefMsgLength = ndefMessage.toByteArray().length;

        // 获取 Ndef 对象
        Ndef ndef = Ndef.get(tag);
        // 只有 Ndef 对象不为空时，才能尝试读取 NDEF 格式数据
        if (ndef == null) {
                contentTv.setText("NFC 标签不支持 NDEF 格式数据！");
                return;
        }
        try {
                // 连接标签
                ndef.connect();
                if (!ndef.isConnected()) {
                        contentTv.setText("连接标签失败！");
                        return;
                }
                // 检查标签是否为只读
                if (!ndef.isWritable()) {
                        contentTv.setText("该标签为只读标签，无法写入数据");
                        return;
                }
                // 检查存储空间
                if (ndef.getMaxSize() < ndefMsgLength) {
                        contentTv.setText("该标签存储空间不足");
                        return;
                }

                // 向标签写入数据
```

```
            ndef.writeNdefMessage(ndefMessage);
            contentTv.setText("已成功写入数据！");
        } catch (Exception e) {
            contentTv.setText("写入数据失败！" + e.getMessage());
            e.printStackTrace();
        } finally {
            try {
                // 关闭与标签的连接
                ndef.close();
            } catch (IOException e) {
                e.printStackTrace();
            }
        }
}
```

上述代码实现了 NDEF 纯文本格式的写入操作。NdefMessage 对象中封装的是 NdefRecord 数组，也就是说，可以创建多个 NdefRecord 对象，写入标签中。为简单易懂起见，本示例中只写入一个 NdefRecord 对象。

至此，向 NDEF 格式标签中写入纯文本数据功能已经实现，运行程序可进行验证。

(9) 实现 readNdef()方法，此方法可读取 NDEF 纯文本数据，代码如下：

```
/**
 * 读取 NDEF 数据
 */
private void readNdef() {
    Ndef ndef = Ndef.get(tag);
    if (ndef == null) {
        contentTv.setText("NFC 标签不支持 NDEF 格式数据！");
        return;
    }
    try {
        ndef.connect();
        if (!ndef.isConnected()) {
            contentTv.setText("连接标签失败！");
            return;
        }
        NdefMessage ndefMsg = ndef.getNdefMessage();
        // 为简单起见，只读取第一个 NdefRecord
        NdefRecord record = ndefMsg.getRecords()[0];
        int contentLength = ndefMsg.toByteArray().length;
        String content = parseToText(record);
        contentTv.setText("数据长度：" + contentLength);
```

```
            contentTv.append("\nNDEF 数据：\n" + content);
        } catch (Exception e) {
            e.printStackTrace();
        }
}
```

上述代码实现了读取 NDEF 纯文本数据：首先获取到 NdefMessage 对象，在此只获取 NdefRecord 数组对象中第一个 NdefRecord 对象；之后通过编写 parseToText()方法，将 NdefRecord 对象解析为文本数据并返回。

(10) 实现 parseToText()方法，此方法作为读取 NDEF 纯文本数据的核心，代码如下：

```
/**
 * 解析 NdefRecord 对象，得到文本数据
 *
 * @param record
 * @return
 */
private String parseToText(NdefRecord record) {
    // NDEF 文本格式的 TNF 必须为 TNF_WELL_KNOWN、RTD 必须为 RTD_TEXT
    if (record.getTnf() != NdefRecord.TNF_WELL_KNOWN) {
        return "TNF 非 TNF_WELL_KNOWN";
    }
    if (!Arrays.equals(record.getType(), NdefRecord.RTD_TEXT)) {
        return "RTD 非 RTD_TEXT";
    }
    try {
        byte[] payload = record.getPayload();
        // 通过"与运算"获取编码方式，0:UTF-8、1:UTF-16
        int utf = (payload[0] & 0x80);
        String contentEncoding = (utf == 0) ? "UTF-8" : "UTF-16";
        // 计算语言编码长度
        int langCodeLength = payload[0] & 0x3f;
        // 获取语言编码
        String langCode = new String(payload, 1, langCodeLength, "US-ASCII");
        String text = new String(payload, langCodeLength + 1,
                    payload.length - langCodeLength - 1, contentEncoding);
        return "语言编码：" + langCode + "\n 内容：" + text;
    } catch (Exception e) {
        e.printStackTrace();
        return "读取数据失败";
    }
}
```

在上述代码中，根据标准 NDEF 纯文本数据格式，将 NdefRecord 对象解析为纯文本数据，并返回。

至此，读写 NDEF 纯文本数据功能已经实现。写入数据如图 7-11 所示，读取数据如图 7-12 所示。

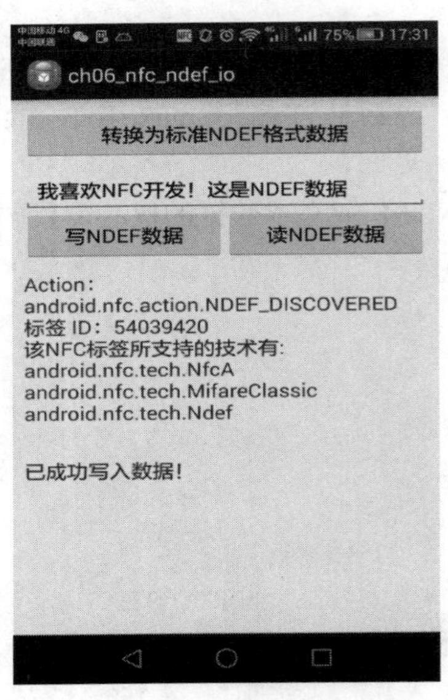

图 7-11　写入数据

图 7-12　读取数据

本 章 小 结

(1) NFC(近场通信技术)是一种非接触式的短距离通信技术，通常有效距离在 4 cm 之内。

(2) NDEF 数据格式的标签可带有 URL、vCard 和 NFC 定义的各种数据类型。

(3) NdefFormatable 数据格式可转换为 NDEF 数据格式。

(4) MifareClassic 数据格式的标签通常有 16 个、32 个或 64 个分区，每个分区有 4 个块(Block)，每个块有 16 个 byte 数据。

(5) 读写 MifareClassic 数据格式的标签时，只有通过验证的分区，才能进行操作。

本 章 练 习

(1) NFC 标签工作模式分为__________和__________。

(2) NDEF 数据被封装到__________对象中，该消息包含了一个或多个__________。

(3) NFC 标签过滤机制分为三个级别，下列级别中，不包含的是__________。

(A) ACTION_NDEF_DISCOVERED　　(B) ACTION_TECH_DISCOVERED

(C) ACTION_NFC_DISCOVERED　　(D) ACTION_TAG_DISCOVERED

(4) 简述 NDEF 数据格式的定义规则。

第 8 章　资源与国际化

本章目标

- 理解 Android 中资源的分类
- 掌握资源的适配
- 理解国际化概念
- 掌握实现国际化的方式

所谓资源，就是 Android 项目中，除了 JAVA 代码部分的其他内容，例如“res”文件夹中的文件都属于资源文件。而这些资源文件中，定义了各种类型的资源，包括程序中使用到的字符串资源、图片资源、布局管理资源等。而国际化(internationalization，I18N)是设计和制造容易适应不同区域要求的产品的一种方式。它要求从产品中抽离所有地域语言，及国家/地区与文化相关的元素。换言之，考虑应用程序在不同地区运行的需要，其功能和代码设计简化了不同本地版本的生产。开发这样的程序的过程，就称为国际化。

Android 国际化的实现与资源密切相关，本章将首先讲解资源的创建与使用，之后讲解 Android 应用程序的国际化技术。

8.1 Android 资源

Android 建议将程序中所有资源外部化，所谓外部化是指将代码中、布局文件中以及其他相关文件中出现的字符串、图片等元素统一提取出来，放入特定的资源文件中，这样有利于单独维护这些资源。

另外，Android 设备屏幕分辨率大小不一，因而，一套资源无法完全匹配所有设备的显示。针对此类情况，Android 做了相应的处理，在开发程序时，为不同的屏幕适配特定的资源，以满足不同需求。这些资源被分组到特定的目录中，当程序运行时，Android 会根据当前设备情况，使用适当的资源。这些资源包括不同的 UI 布局、不同的语言显示等。

8.1.1 Android 资源概述

Android 程序开发时，除了默认资源外，通常还会创建多组资源，用于对主流设备屏幕的匹配以及不同地区语言的适配。这些资源也称为备用资源，在程序运行时，Android 会自动检测当前设备的配置情况，为程序加载合适的资源。

1．资源的分类

Android 项目中有两个目录可以存放资源文件，分别是“assets”和“res”目录，两者的区别在于：

- ✧ “assets”目录通常用于存放普通文件资源，这些文件将不做任何改动地被添加到程序安装包中，使用这部分资源时，可以通过文件名访问。
- ✧ “res”目录中的文件资源会自动在项目中的“R.java”类(gen/包名/R.java)中引用，并生成唯一的 ID 供程序使用。

这两个目录中的资源使用方式将在接下来的小节中详细介绍，本小节主要介绍“res”目录中的资源。

Android 对“res”目录中的资源进行了分类，将不同类别的资源放入特定的目录中，新建 Android 4.0 项目后，打开“res”目录，结构如图 8-1 所示。

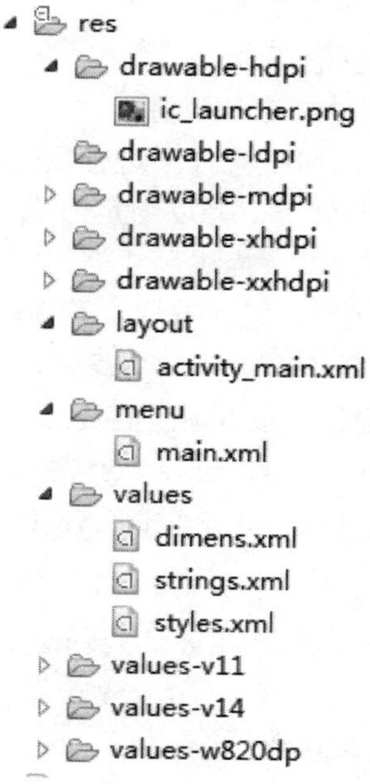

```
res
  drawable-hdpi
    ic_launcher.png
  drawable-ldpi
  drawable-mdpi
  drawable-xhdpi
  drawable-xxhdpi
  layout
    activity_main.xml
  menu
    main.xml
  values
    dimens.xml
    strings.xml
    styles.xml
  values-v11
  values-v14
  values-w820dp
```

图 8-1　res 目录结构

“res”目录中包含所有程序中用到的资源(上图中只是部分资源目录)，这些资源又被放到不同的子目录中，其中分为：“drawable”目录、“layout”目录、“menu”目录和“values”目录。这些资源所在的目录名称非常重要，必须按照一定的格式定义。需要注意的是，这些目录只是其中一部分。“res”目录中支持的资源子目录如下：

- ✧ res/animator：存放定义属性动画的 XML 文件。
- ✧ res/anim：存放定义渐变动画的 XML 文件(animator 目录中的文件也可以放到此目录下，但通常不建议这样做)。
- ✧ res/color：存放定义颜色状态列表的 XML。
- ✧ res/drawable：存放图片文件(.jpg、.png、.9.png、.gif)，或者编译为 Drawable 资源子类型的 XML 文件，例如：定义图片、形状、状态列表，或其他 Drawable 资源。
- ✧ res/mipmap：适应不同程序图标密度的 Drawable 文件，通常不需要进行特殊设置。
- ✧ res/layout：存放定义 UI 布局的 XML 文件。
- ✧ res/menu：存放定义程序菜单(选项菜单、上下文菜单等)的 XML 文件。
- ✧ res/raw：以原始形式保存的任意文件，通过使用资源 ID 获取文件流，如果需要访问原始文件名和文件层次结构，就需要将这部分资源放到“assets”目录中。
- ✧ res/values：存放字符串、数值和颜色等简单值的 XML 文件。与其他子目录不同的是，此目录中的 XML 文件可以描述多个资源，根节点为<resources>元素。每个子元素定义一种资源，例如：

✧ <string>元素：用于创建字符串资源。

✧ <color>元素：用于创建颜色值资源。

✧ <dimen>元素：用于描述尺寸类数值资源。

虽然可以把这些资源放在任意 XML 文件中进行定义，但为了增加程序的易读性，通常将不同的资源分在不同的 XML 文件下，并以约定俗成的格式给文件命名。例如：

✧ dimens.xml：保存<dimen>元素，描述尺寸类数值资源。

✧ strings.xml：保存<string>元素，描述普通字符串资源。

✧ styles.xml：保存<style>元素，描述样式类资源。

✧ colors.xml：保存<color>元素，描述颜色值资源。

✧ arrays.xml：保存<string-array><integer-array>等元素，描述数组资源。

✧ res/xml：存放任意 XML 文件，在这里的 XML 文件通常是一些配置文件或存储数据的文件。

注 意

因为在 R.java 类中引用了 res 目录中所有的资源 ID，所以如果 R.java 类“消失”，请首先检查 res 目录中的资源格式是否出现错误，通常表现为文件名不规范、引用资源不存在等。Android 规定资源文件的命名只能由小写字母(a ~ z)、数字(0 ~ 9)、下划线(_)和点(.)组成。

2. 资源限定符

通俗地说，资源限定符就是用于区分不同备用资源的标识，Android 会根据设备的配置情况查找匹配的某组资源来引用。

开发者可以在“res”目录中新建不同的子目录，把满足不同条件的配置资源放到这些目录中，目录的命名方式有一定的规范：文件夹名称以“drawable”开头，表示资源类别的名称；中间用中划线(“-”)分隔；中划线之后的部分为限定符，限定符可以追加多个，但必须按照一定规则定义，如果有错误，该目录中的资源将被忽略。完整的资源目录命名方式有：“drawable”“drawable-en-hdpi”“values-en”等。

Android 支持若干限定符的配置，这些限定符在使用时，必须按照一定顺序设置。Android 中常见资源限定符如表 8-1 所示。

表 8-1　Android 中常见资源限定符

限定符名称	描　　述	示　　例
语言和区域	语言限定符是由两个字母组成的 ISO 639-1 语言代码定义，之后可追加两个字母组成的 ISO 3166-1-alpha-2 区域码，区域码前需要添加一个小写的“r”，区域码是可选的 中文：zh 英语：en 中文简体：zh-rCN 英国英语：en-rUS	匹配中文语言，资源文件夹命名为： “res/values-zh”

续表

限定符名称	描　述	示　例
smallestWidth	用于匹配设备的可用屏幕最窄边的尺寸，格式为 sw<N>dp，dp 为单位，<N>为数值	如果布局要求设备可用屏幕最窄边至少为 480dp，那么布局资源文件夹可命名为： “res/layout-sw480dp/”。 当资源目录中提供了多个 smallestWidth 限定符值时，Android 会使用最接近(但不超出)的值
可用宽度	用于匹配设备的可用屏幕最小宽度，格式为 w<N>dp，在横向和纵向屏幕切换时，为了匹配实际屏幕宽度，此配置也会自动变化	如果布局要求设备可用屏幕的最小宽度是 480dp，那么布局资源文件夹可命名为： “res/layout-w480dp/”
可用高度	用于匹配设备的可用屏幕最小高度，格式为 h<N>dp，在横向和纵向屏幕切换时，为了匹配实际屏幕宽度，此配置也会自动变化	如果布局要求设备可用屏幕的最小高度是 800dp，那么布局资源文件夹可命名为： “res/layout-h800dp/”
屏幕方向	用于匹配屏幕方向，在横向和纵向屏幕切换时，可能会发生变化。 port：纵向屏幕 land：横向屏幕	如果布局匹配横向屏幕，并且是高密度屏幕，布局资源文件夹可命名为： “drawable-port-hdpi”
夜间模式	该模式会根据当天时间进行调整，也可进行手动设置。 night：夜间 notnight：白天	匹配夜间模式，布局资源文件夹可命名为： “drawable-night”
屏幕像素密度	用于匹配不同设备屏幕的密度。 ldpi：低密度，约为 120dpi mdpi：中等密度，约为 160dpi hdpi：高密度，约为 240dpi xhdpi：超高密度，约为 320dpi 需要注意的是，如果有更符合屏幕配置的资源，Android 会做出自动选择	匹配高密度(240dpi)屏幕，布局资源文件夹可命名为： “drawable-hdpi”
触摸屏类型	匹配是否有触摸屏 notouch：设备没有触摸屏 finger：设备有一个可用手指直接交互的触摸屏	匹配无触摸屏的设备，布局资源文件夹可命名为： “drawable-finger”

8.1.2 资源的创建与使用

了解了资源的分类与限定符后，接下来将通过代码示例来讲解常见资源的创建与使用。在此之前，首先需要了解 Android 中的 Resources 类。

Resources 类位于 android.content.res 包中，用于在 JAVA 代码中访问应用程序资源。在 Activity 中，可以直接通过 getResources()方法获取 Resources 对象。Resources 类常用方法如表 8-2 所示。

表 8-2 Resources 类常用方法

方 法 名	描 述
getAnimation(int id)	获取动画资源，返回值为 XmlResourceParser 对象
getAssets()	获取 AssetManager 对象，用于操作 assets 文件夹中的资源
getColor(int id)	获取<color>资源
getDimension(int id)	获取<dimen>资源
getDimensionPixelSize(int id)	获取<dimen>资源作为原始像素
getDrawable(int id)	获取 Drawable 资源
getIntArray(int id)	获取<integer-array>资源
getString(int id)	获取<string>资源
getStringArray(int id)	获取<string-array>资源
getXml(int id)	获取 XML 文件资源，返回值为 XmlResourceParser 对象
getColorStateList(int id)	获取颜色状态列表资源，返回值为 ColorStateList 对象

1．res/color 目录中的资源

该目录用于存放定义颜色状态列表(ColorStateList)的 XML 文件资源，可以根据 View 的状态改变 View 的内容颜色。View 的状态包括：按下、选中、获取焦点、不可用等。例如，可以设置一个按钮在不同状态下文字的颜色。

定义颜色状态列表的基本格式如下：

```
<?xml version="1.0" encoding="utf-8"?>
<selector xmlns:android="http://schemas.android.com/apk/res/android" >
    <item
        android:color="hex_color"
        android:state_pressed=["true" | "false"]
        android:state_focused=["true" | "false"]
        android:state_selected=["true" | "false"]
        android:state_checkable=["true" | "false"]
        android:state_checked=["true" | "false"]
        android:state_enabled=["true" | "false"]
        android:state_window_focused=["true" | "false"] />
    <item … />
    ……
```

```
</selector>
```

XML 文件最外层是<selector>元素，其子元素由一个或多个<item>组成，<item>元素中定义了不同的属性：

✧ color：十六进制的颜色值，在 XML 文件资源中使用的颜色值是一个十六进制的 RGB 值，并且支持 Alpha 通道。定义颜色的格式为：以“#”开头，后面紧接着 Alpha-Red-Green-Blue，例如：#RGB、#ARGB、#RRGGBB、#AARRGGBB。之后不再做介绍。

✧ state_pressed：按下的状态，取值为 boolean 类型。

✧ state_focused：是否获取到焦点，取值为 boolean 类型。

✧ state_selected：是否已选择，取值为 boolean 类型。

✧ state_checkable：是否可选，取值为 boolean 类型。

✧ android:state_checked：是否被选中，常见用于复选框、单选框等，取值为 boolean 类型。

✧ state_enabled：是否可用，取值为 boolean 类型。

✧ state_window_focused：程序是否已获得焦点(是否在当前屏幕)，取值为 boolean 类型。

值得注意的是，这些属性并不适用于所有控件，在使用之前，应当按照实际需求进行选择。

(1) 在 res/color 目录中创建资源文件“btn_text_color_status.xml”，用于设置按钮在按下或默认状态下文字颜色的变化，代码如下：

```
<?xml version="1.0" encoding="utf-8"?>
<selector xmlns:android="http://schemas.android.com/apk/res/android">
    <!--按下状态 -->
    <item
       android:state_pressed="true"
       android:color="#ff0000"/>
    <!--默认状态 -->
    <item android:color="#000000"/>
</selector>
```

(2) 在程序中使用颜色状态列表资源。

✧ 在 Java 代码中：

```
Button mButton= (Button) findViewById(R.id.act_main_btn);
ColorStateListcolorId =getResources().getColorStateList(R.color.btn_text_color_status);
mButton.setTextColor(colorId);
```

✧ 在 XML 文件中：

```
<Button
    android:layout_width="wrap_content"
    android:layout_height="wrap_content"
    android:text="我喜欢做 Android 开发"
```

```
    android:textColor="@color/btn_text_color_status" />
```

2. res/drawable 目录中的资源

该目录中存放各种图片资源和可以转为 Drawable 资源的 XML 文件。对于单张图片资源的使用，操作比较简单，不再过多介绍。接下来重点讲解一下图片状态列表资源的创建及使用。

图片状态列表(DrawableStateList)资源类似于颜色状态列表资源，区别在于，前者是设置不同状态的图片而不是颜色。定义图片状态列表的格式基本与定义颜色状态列表的格式相同：

```
<?xml version="1.0" encoding="utf-8"?>
<selector xmlns:android="http://schemas.android.com/apk/res/android"
    android:constantSize=["true" | "false"]
    android:dither=["true" | "false"]
    android:variablePadding=["true" | "false"] >
    <item
        android:drawable="@[package:]drawable/drawable_resource"
        android:state_pressed=["true" | "false"]
        android:state_focused=["true" | "false"]
        android:state_hovered=["true" | "false"]
        android:state_selected=["true" | "false"]
        android:state_checkable=["true" | "false"]
        android:state_checked=["true" | "false"]
        android:state_enabled=["true" | "false"]
        android:state_activated=["true" | "false"]
        android:state_window_focused=["true" | "false"] />
</selector>
```

<selector>元素中的属性解释如下：

- ✧ constantSize：是否根据当前状态改变尺寸，取值为 boolean 类型，默认为 false。
- ✧ dither：是否支持图像的抖动处理，当颜色值低于 8 位时，可以保持较好的显示效果，取值为 boolean 类型，默认为 true。
- ✧ variablePadding：图像内边距的尺寸是否根据当前状态改变，取值为 boolean 类型，默认为 false。

<item>元素中的属性与颜色状态列表中<item>元素属性基本相同，其中：

- ✧ drawable：图像资源，此属性是必需的。
- ✧ state_ hovered：目标对象被按下又弹起后的状态，取值为 boolean 类型。
- ✧ state_activated：目标处于活动后的状态，取值为 boolean 类型。

(1) 在 res/drawable 目录中创建资源文件“btn_background_status.xml”，用于设置按钮在按下或默认状态下背景图片的变化，代码如下：

```
<?xml version="1.0" encoding="utf-8"?>
```

```
<selector xmlns:android="http://schemas.android.com/apk/res/android">
    <!-- 按下状态 -->
    <item
        android:drawable="@drawable/btn_pressed"
        android:state_pressed="true"/>
    <!-- 默认状态 -->
    <itemandroid:drawable="@drawable/btn_normal"/>
</selector>
```

(2) 在程序中使用图片状态列表资源。

✧ 在 Java 代码中：

```
mButton.setBackgroundResource(R.drawable.btn_background_status);
```

需要注意的是，对 Button 对象设置背景资源时，可以直接把资源 ID 传入 setBackgroundResource()方法中进行设置，不需要再调用 getResources()方法获取 Resources 对象。

✧ 在 XML 文件中：

```
<Button
    android:layout_width="wrap_content"
    android:layout_height="wrap_content"
    android:background="@drawable/btn_background_status"
    android:text="我喜欢做 Android 开发" />
```

3. res/values 目录中的<dimen>尺寸资源

<dimen>元素用于描述程序中用到的尺寸等度量单位数值，例如：5 dp、10 px、16 sp 等。Android 支持的度量单位有：

- ✧ dp(dip)：设备独立像素，该单位不依赖于像素，表示每英寸有多少个显示点，根据屏幕密度计算实际长度数值，是 Android 推荐的长度单位。
- ✧ sp：与比例无关的像素，类似于 dp，通常用于描述字体大小，也是 Android 推荐的用于描述字体大小的单位。
- ✧ pt：point，点，1pt＝1/72 英寸(in)，基于屏幕的物理尺寸。
- ✧ px：屏幕实际像素点，通常不建议使用此单位描述 UI 资源长度。
- ✧ mm：毫米，基于屏幕的物理尺寸。
- ✧ in：英寸，基于屏幕的物理尺寸。

(1) 使用<dimen>元素定义数值，文件位置为“res/values/dimens.xml”，代码如下：

```
<?xml version="1.0" encoding="utf-8"?>
<resources>
    <dimen name="btn_width">260dp</dimen>
    <dimen name="btn_text_size">20sp</dimen>
</resources>
```

代码中定义了两个按钮属性的数值：按钮宽度为 260 dp，按钮文字大小为 20 sp。

(2) 在程序中使用<dimen>元素定义数值。

✧ 在 Java 代码中：

```
float width = getResources().getDimension(R.dimen.btn_width);
float textSize=getResources().getDimension(R.dimen.btn_textsize);
mButton.setWidth((int) width);
mButton.setTextSize(TypedValue.COMPLEX_UNIT_PX,textSize);
```

✧ 在 XML 文件中：

```
<Button
    android:layout_width="@dimen/btn_width"
    android:layout_height="wrap_content"
    android:text="我喜欢做 Android 开发"
    android:textSize="@dimen/btn_text_size" />
```

4．res/values 目录中的<string>字符串资源

<string>元素用于程序中字符串文本的引用，可方便地实现 Android 程序的国际化功能，关于 Android 的国际化问题，在后续章节中会一一进行介绍。

(1) 使用<string>元素定义字符串，文件位置为“res/values/strings.xml”，代码如下：

```
<?xml version="1.0" encoding="utf-8"?>
<resources>
    <string name="btn_name_hello">我喜欢做 Android 开发</string>
</resources>
```

(2) 在程序中使用<string>元素定义字符串。

✧ 在 Java 代码中：

```
String str=getResources().getString(R.string.btn_name_hello);
```

✧ 在 XML 文件中：

```
<Button
    android:layout_width="wrap_content "
    android:layout_height="wrap_content"
    android:text="@string/btn_name_hello" />
```

5．res/values 目录中的<color>颜色值资源

<color>元素用于描述十六进制的颜色值。

(1) 使用<color>元素定义颜色值，文件位置为“res/values/colors.xml”，代码如下：

```
<?xml version="1.0" encoding="utf-8"?>
<resources>
    <color name="btn_text_color">#00ff00</color>
</resources>
```

(2) 在程序中使用<color>元素定义颜色值。

✧ 在 Java 代码中：

```
intcolorId = getResources().getColor(R.color.btn_text_color);
mButton.setTextColor(colorId);
```

✧ 在 XML 文件中：

```
<Button
    android:layout_width="wrap_content"
    android:layout_height="wrap_content"
    android:text="我喜欢做 Android 开发"
    android:textColor="@color/btn_text_color"/>
```

6．res/xml 目录中的资源

该目录中存放有各种 XML 文件，包括配置相关的 XML 文件，这些 XML 资源文件只在代码程序中使用，而不会用于其他 XML 文件。通过 Java 代码获取 res/xml 目录中的资源：

```
XmlResourceParser xml = getResources().getXml(R.xml.xml_data);
```

8.2　国际化

安装有 Android 系统的设备遍布全球各地，为了能够满足更多用户的需求，一款应用程序通常需要根据目标用户群体，设置不同的语言版本，如对文本、音频文件、数字、货币和图片等资源的配置，以适应不用地区用户的使用需求。简单地说，Android 国际化就是应用程序会根据不同地区的用户设备设置不同的语言，自动显示不同资源版本的界面。

对开发人员而言，在 Android 程序中实现国际化配置比较简单，只需在项目的资源目录中，按照特定的规范与实际需求添加用于适配不同语言的文件资源即可。由此，开发人员不需要为目标用户群体开发不同语言版本的程序，大大降低了开发成本，也避免了资源的浪费。

Android 中实现程序的国际化大致可分为跟随系统国际化、程序内国际化两种类型。

注　意

若想了解不同国家和地区的语言代码请参见本书最后的“附录 国家地区语言代码表”，需要注意的是资源的命名规范，地区代码之前需要加入小写英文字母“r”。

8.2.1　跟随系统国际化

所谓跟随系统国际化，就是程序根据系统所设置的语言或地区，由系统选择合适的备用资源，前提是开发者已经对该语言添加了备用资源，如果系统找不到合适资源，将使用默认资源。跟随系统国际化的实现方式比较简单，几乎不需要编写代码，只需添加正确的备用资源就可实现。

1．语言资源的配置

Android 项目中，语言资源通常被放置于“res/values/strings.xml”文件中，当然，XML 文件名字可以根据需求自行确定。实现语言国际化就是根据不同国家或地区语言，创建不同的“values”目录，目录名称通过资源限定符区分。

下述示例用于实现：通过创建不同的语言资源，完成根据系统当前语言实现国际化的功能，要求匹配的语言有中文简体、中文繁体、英文。

(1) 创建项目“ch08_i18n_ sys”，并添加语言相关备用资源。

✧ 修改“res/values/strings.xml”文件，代码如下：

```
<resources>
    <string name="app_name">ch08_i18n_sys</string>
    <string name="act_main_crt_lang">当前语言为：中文简体(默认)</string>
</resources>
```

✧ 创建资源文件“res/values-en/strings.xml”，用于匹配英文环境，代码如下：

```
<?xml version="1.0" encoding="utf-8"?>
<resources>
    <string name="act_main_crt_lang">Current Language：English</string>
</resources>
```

✧ 创建资源文件“res/values-zh-rCN/strings.xml”，用于匹配简体中文环境，代码如下：

```
<?xml version="1.0" encoding="utf-8"?>
<resources>
    <string name="act_main_crt_lang">当前语言为：中文简体</string>
</resources>
```

✧ 创建资源文件“res/values-zh-rTW/strings.xml”，用于匹配繁体中文环境，代码如下：

```
<?xml version="1.0" encoding="utf-8"?>
<resources>
    <string name="act_main_crt_lang">當前語言為：中文繁體</string>
</resources>
```

在上述这些文件中，“res/values/strings.xml”为默认资源文件，其他三个文件为备用资源文件，系统在备用资源中找不到的元素，会自动加载到默认资源中，例如默认资源中的“app_name”元素，在其他三个备用资源文件中是没有被定义的。

(2) 修改“activity_main.xml”布局文件，代码如下：

```
<LinearLayoutxmlns:android="http://schemas.android.com/apk/res/android"
    xmlns:tools="http://schemas.android.com/tools"
    android:layout_width="match_parent"
    android:layout_height="match_parent"
    android:background="#ededed"
    android:orientation="vertical"
    android:padding="10dp" >
    <TextView
        android:layout_width="wrap_content"
        android:layout_height="wrap_content"
        android:layout_marginTop="10dp"
        android:text="@string/act_main_crt_lang"
        android:textSize="18sp" />
```

```
</LinearLayout>
```

在上述代码中，TextView 控件的 text 属性设置的文本内容引用了资源文件中的“act_main_crt_lang”元素。

(3) 运行程序。在手机的系统设置里修改语言后，再次运行本程序，观察语言变化。结果如图 8-2 所示。

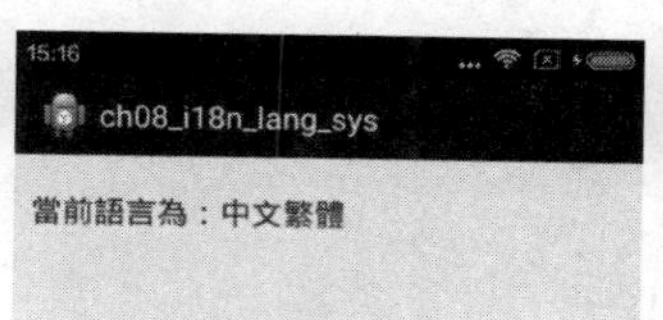

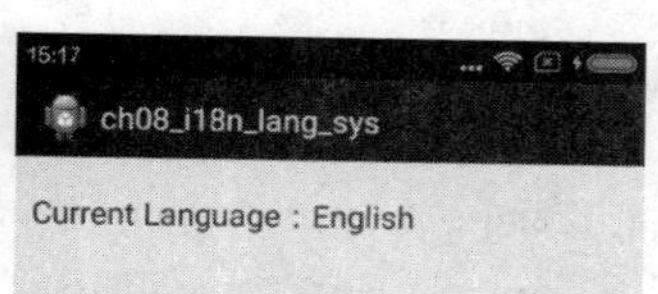

图 8-2　语言资源适配效果

“res/values”资源目录中除了“string”资源，还有其他资源，使用方法与“string”资源相同，读者可以自行验证。

注意

备用资源的名称必须和默认资源名称相同，如果不同，将被视为两种不同的资源，可能不会被程序使用到。

2．Drawable 资源的适配

Drawable 资源被放置于“res/drawable”资源目录中，通过创建不同资源限定符的子目录也可以实现国际化功能。

下述示例用于实现：通过创建不同 Drawable 资源目录，实现图片的替换，当系统语言为中文时，显示中国国旗图片；当系统语言为英文时，显示英国国旗图片。

(1) 在“ch08_i18n_ sys”项目基础上，添加 Drawable 资源目录。

- ✧ 将“中国国旗图片”(名称为 icon_country.png)拷贝到“res/drawable-hdpi”目录下，作为默认国旗图片。
- ✧ 创建“res/drawable-zh-hdpi”目录，用于适配中文环境，并将“中国国旗图片”拷贝到此目录下。
- ✧ 创建“res/drawable-en-hdpi”目录，用于适配英文环境，并将“英国国旗图片”(名称同为 icon_country.png)拷贝到该目录下。

至此，Drawable 资源已经创建完毕。

(2) 修改“activity_main.xml”布局文件，在原基础上添加一个 ImageView 控件，用于显示国旗图片，代码如下：

```
<LinearLayoutxmlns:android="http://schemas.android.com/apk/res/android"
    xmlns:tools="http://schemas.android.com/tools"
    android:layout_width="match_parent"
    android:layout_height="match_parent"
    android:background="#ededed"
    android:orientation="vertical"
    android:padding="10dp" >

    <ImageView
```

```
        android:id="@+id/imageView1"
        android:layout_width="wrap_content"
        android:layout_height="wrap_content"
        android:layout_gravity="center_horizontal"
        android:src="@drawable/icon_country" />
    <TextView
        android:layout_width="wrap_content"
        android:layout_height="wrap_content"
        android:layout_marginTop="10dp"
        android:text="@string/act_main_crt_lang"
        android:textSize="18sp" />
</LinearLayout>
```

(3) 运行程序后，在手机的系统设置里进行中英文切换，观察程序中国旗图片的变化。显示结果如图 8-3 所示。

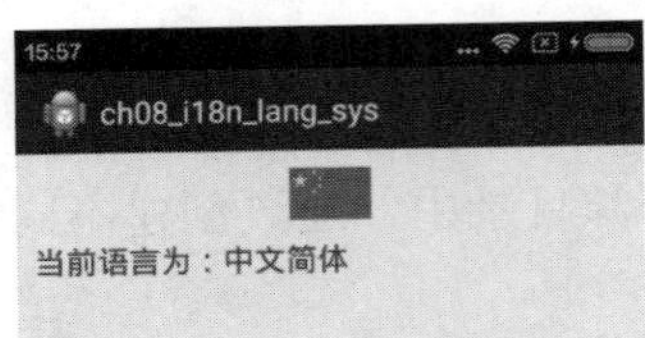

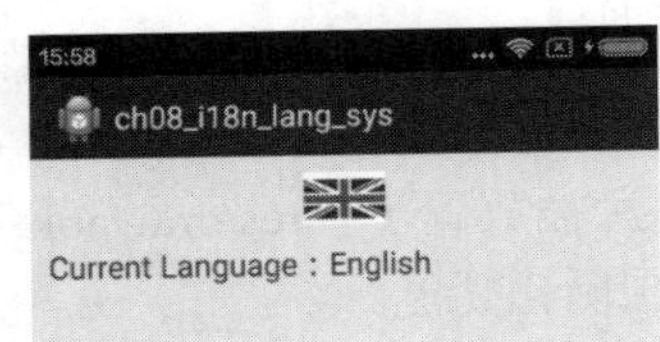

图 8-3　Drawable 资源的适配效果

3. Layout 布局资源的适配

Layout 布局资源的国际化意义在于，不同的国家或地区可能对程序界面要求不同，有的可能要求从左到右阅读，有的可能要求从右到左阅读。

Layout 布局资源被放置于“res/layout”资源目录中，通过创建不同资源限定符的子目录也可以实现国际化功能。

下述示例用于实现：通过修改“ch08_i18n_ sys”项目，实现当系统语言为中文时，国家图片位于文本的上方；当语言为英文时，国家图片位于文本的下方。

(1) 创建“res/layout-en”资源目录，将“res/layout”目录中已存在的“activity_main.xml”布局文件复制到该目录下，并修改代码：

```
<LinearLayoutxmlns:android="http://schemas.android.com/apk/res/android"
    xmlns:tools="http://schemas.android.com/tools"
    android:layout_width="match_parent"
    android:layout_height="match_parent"
    android:background="#ededed"
    android:orientation="vertical"
    android:padding="10dp" >
    <TextView
        android:layout_width="wrap_content"
        android:layout_height="wrap_content"
```

```
        android:layout_marginTop="10dp"
        android:text="@string/act_main_crt_lang"
        android:textSize="18sp" />
    <ImageView
        android:id="@+id/imageView1"
        android:layout_width="wrap_content"
        android:layout_height="wrap_content"
        android:layout_gravity="center_horizontal"
        android:src="@drawable/icon_country" />
</LinearLayout>
```

上述代码仅将 ImageView 控件改为放置于 TextView 控件下方。

(2) 中文资源不再创建新的 Layout 资源，将直接使用默认资源。

(3) 运行程序后，在手机的系统设置里进行中英文切换，观察在不同语言环境下布局的变化，运行结果如图 8-4 所示。

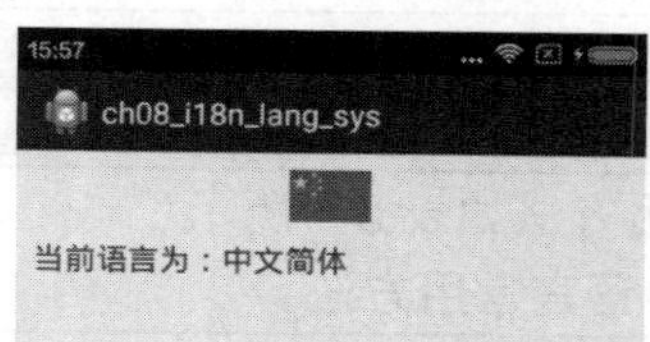

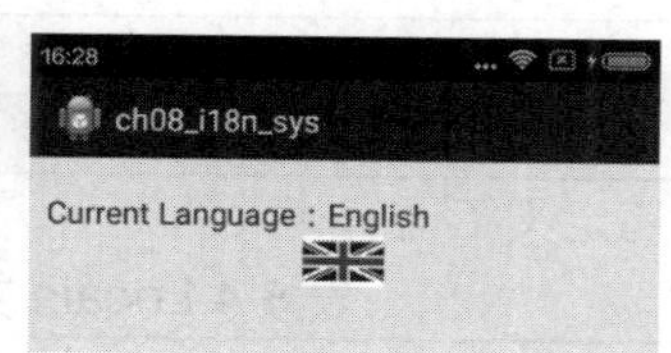

图 8-4　Layout 布局资源的适配效果

8.2.2　程序内国际化

所谓程序内国际化，就是程序中的语言环境不再根据系统所设置的语言或地区的不同而发生改变，而是由程序本身来设定自身的语言环境，程序可以提供给用户所有支持的语言，由用户来选择所需语言。实现程序内国际化并不影响系统本身的语言环境，甚至可以设置系统本身所不存在的语言环境，例如，小米 MIUI 系统中仅支持中文与英文的语言环境，程序内国际化的实现，可以为用户提供其他语言环境。

实现程序内国际化，需要用到一个比较重要的类：Locale 类。该类位于“java.util”包中，用于描述不同的国家及地区，操作系统的相关语言等。该类中定义了一系列用于描述国家或地区语言的常量，这些常量为 Locale 类型，如表 8-3 所示。

表 8-3　Locale 类中语言相关常量

常　量　名	语言及编码
ROOT	当前系统语言
CANADA	加拿大英语，表示国家或地区，en_CA
CANADA_FRENCH	加拿大法语，表示国家或地区，fr_CA
CHINA	中文简体，表示国家或地区，zh_CN
CHINESE	中文，表示语言，zh
SIMPLIFIED_CHINESE	中文简体，表示语言，zh_CN

续表

常　量　名	语言及编码
TRADITIONAL_CHINESE	中文繁体，表示语言，zh_TW
ENGLISH	英语，表示语言，en
FRANCE	法语，表示国家或地区，fr_FR
FRENCH	法语，表示语言，fr
GERMAN	德语，表示语言，de
GERMANY	德语，表示国家或地区，de_DE
ITALIAN	意大利语，表示语言，it
ITALY	意大利语，表示国家或地区，it_IT
JAPAN	日语，表示国家或地区，ja_JP
JAPANESE	日语，表示语言，ja
KOREA	韩国语，表示国家或地区，ko_KR
KOREAN	韩国语，表示语言，ko
UK	英国英语，表示国家或地区，en_GB
US	美国英语，表示国家或地区，en_US

Locale 类中还定义了语言相关的方法，常用方法如表 8-4 所示。

表 8-4 Locale 类中语言常用方法

常　量　名	语言及编码
Locale(String language)	构造方法，根据语言代码创建 Locale 对象
Locale(String language, String country)	构造方法，根据语言、国家或地区代码创建 Locale 对象
Locale(String language, String country, String variant)	构造方法，根据语言、国家或地区、变量代码创建 Locale 对象
getAvailableLocales()	获取所有已安装语言，返回值为 Locale 数组
getLanguage()	获取当前语言环境的语言编码
getDefault()	获取当前默认语言环境，返回值为 Locale 对象
getCountry()	获取当前语言环境的国家或地区编码
getVariant()	获取当前语言环境的变量编码
getDisplayCountry()	获取适合当前用户语言环境的国家或地区编码
getDisplayLanguage()	获取适合当前用户语言环境的语言编码
getDisplayName()	获取适合当前用户的语言环境名称
getDisplayVariant()	获取适合当前用户语言环境的变量代码名称
setDefault(Locale newLocale)	静态方法，设置当前默认语言环境

实现程序内国际化，除了 Locale 类，还需要借助 Resources 类，通过该类的 updateConfiguration(Configuration config, DisplayMetrics metrics)方法更新程序资源配置，才能最终实现程序内国际化。程序内国际化的实现存在以下需要解决的问题：

✧ 在设置程序内国际化时，之前已经被打开的 Activity 不会自动更新资源，只有重新打开(重新执行 onCreate()方法)，才能实现改变，因此，通常会通过重启相

应的 Activity 或者重启程序解决此问题。

✧ 在程序完全关闭并再次打开时，程序内国际化设置可能会失效，因此，通常会在设置语言时，将设置参数保存到 SharedPreferences 中，每次启动程序时，首先读取该参数，进行自动配置。

下述示例用于实现：通过创建不同的语言资源，实现程序内国际化功能，要求匹配的语言有韩国语和法语。当在首页中单击语言按钮时，打开新的页面并显示相应的语言及国家图片资源。

(1) 创建项目“ch08_i18n_app”，并添加语言相关资源。

✧ 修改“res/values/strings.xml”文件，代码如下：

```
<?xml version="1.0" encoding="utf-8"?>
<resources>
    <string name="app_name">ch08_i18n_app</string>
    <string name="act_main_btn_ko">韩语</string>
    <string name="act_main_btn_fr">法语</string>
    <string name="act_second_hello">当前语言：中文(默认)</string>
</resources>
```

✧ 创建“res/values-fr/ strings.xml”文件，用于匹配法语，代码如下：

```
<?xml version="1.0" encoding="utf-8"?>
<resources>
      <string name="act_second_hello">Languesparlées: françaisactuels</string>
</resources>
```

✧ 创建“res/values-ko/ strings.xml”文件，用于匹配韩语，代码如下：

```
<?xml version="1.0" encoding="utf-8"?>
<resources>
    <string name="act_second_hello">현재언어: 한국어</string>
</resources>
```

以上创建的语言资源中，仅对“act_second_hello”元素进行国际化匹配。

(2) 添加 Drawable 相关资源。

✧ 将“中国国旗图片”(名称为 icon_country.png)拷贝到“res/drawable-hdpi”目录下，作为默认国旗图片。

✧ 将“法国国旗图片”(名称为 icon_country.png)拷贝到“res/drawable-fr-hdpi”目录下。

✧ 将“韩国国旗图片”(名称为 icon_country.png)拷贝到“res/drawable-ko-hdpi”目录下。

(3) 修改“res/layout/activity_main.xml”布局文件，代码如下：

```
<LinearLayoutxmlns:android="http://schemas.android.com/apk/res/android"
    xmlns:tools="http://schemas.android.com/tools"
    android:layout_width="match_parent"
    android:layout_height="match_parent"
```

```
    android:background="#ededed"
    android:orientation="vertical"
    android:padding="10dp" >

    <Button
        android:id="@+id/act_main_ko_btn"
        android:layout_width="fill_parent"
        android:layout_height="wrap_content"
        android:layout_marginTop="10dp"
        android:text="@string/act_main_btn_ko" />
    <Button
        android:id="@+id/act_main_fr_btn"
        android:layout_width="fill_parent"
        android:layout_height="wrap_content"
        android:layout_marginTop="10dp"
        android:text="@string/act_main_btn_fr"
        android:textSize="18sp" />
</LinearLayout>
```

(4) 创建“res/layout/act_second.xml”布局文件，代码如下：

```
<LinearLayoutxmlns:android="http://schemas.android.com/apk/res/android"
    xmlns:tools="http://schemas.android.com/tools"
    android:layout_width="match_parent"
    android:layout_height="match_parent"
    android:background="#ededed"
    android:orientation="vertical"
    android:padding="10dp" >

    <ImageView
        android:layout_width="wrap_content"
        android:layout_height="wrap_content"
        android:layout_gravity="center_horizontal"
        android:src="@drawable/icon_country" />
    <TextView
        android:id="@+id/act_main_info_tv"
        android:layout_width="wrap_content"
        android:layout_height="wrap_content"
        android:layout_marginTop="10dp"
        android:text="@string/act_second_hello"
        android:textSize="18sp" />
</LinearLayout>
```

(5) 创建“SecondActivity.java”类，作为第二个页面，该类中只需要关联“res/layout/act_second.xml”布局文件即可，此处代码省略。

(6) 在“AndroidManifest.xml”文件中声明“SecondActivity.java”类，声明部分的代码如下：

```
<application>
        ……
        <activity android:name=".SecondActivity" />
</application>
```

(7) 修改“MainActivity.java”类，代码如下：

```
public class MainActivity extends Activity {

        /** 韩国语按钮 */
        private Button koBtn = null;
        /** 法语按钮 */
        private Button frBtn = null;

        @Override
        protected void onCreate(Bundle savedInstanceState) {
                super.onCreate(savedInstanceState);
                setContentView(R.layout.activity_main);

                koBtn = (Button) findViewById(R.id.act_main_ko_btn);
                frBtn = (Button) findViewById(R.id.act_main_fr_btn);

                koBtn.setOnClickListener(onBtnClickListener);
                frBtn.setOnClickListener(onBtnClickListener);
        }

        privateOnClickListeneronBtnClickListener = new OnClickListener() {

                @Override
                public void onClick(View view) {
                        if (view == koBtn) {
                                changeAppLanguage(Locale.KOREA);
                        } else if (view == frBtn) {
                                changeAppLanguage(Locale.FRENCH);
                        }
                        Intent intent =
                                new Intent(MainActivity.this, SecondActivity.class);
                        startActivity(intent);
```

```
            }
        };

        /**
         * 修改程序语言
         *
         * @paramlocale
         */
        public void changeAppLanguage(Locale locale) {
                Resources resources = getResources();
                // 获取资源配置对象
                Configuration config = getResources().getConfiguration();
                // 获取屏幕参数
                DisplayMetricsdm = resources.getDisplayMetrics();
                // 设置语言
                config.locale = locale;
                // 更新程序资源配置
                resources.updateConfiguration(config, dm);
        }
}
```

该类中，核心代码是 changeAppLanguage(Locale locale)方法，通过传入的 Locale 对象，可修改当前程序语言环境。

(8) 运行程序，通过单击不同按钮可以改变程序内语言环境，效果如图 8-5 所示。

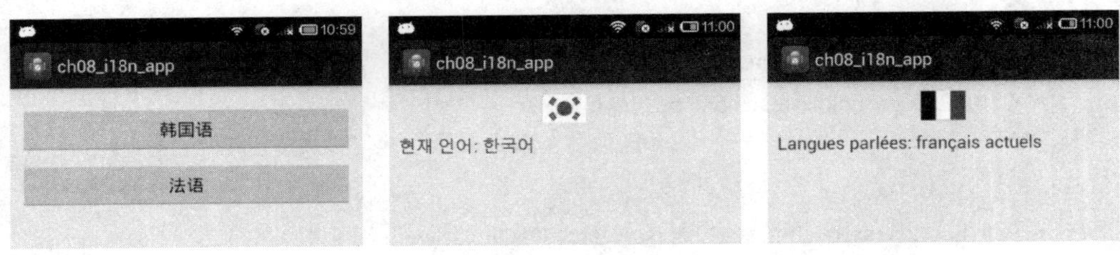

图 8-5　程序内国际化

本 章 小 结

(1) 根据功能和使用方式不同，Android 中的资源被放置于“assets”和“res”目录下。

(2) “res”目录中的资源会自动在项目中的“R.java”类(gen/包名/R.java)中被引用，并生成唯一的 ID 供程序使用。

(3) 程序中通过 Resources 对象获取“res”目录中的资源。

(4) 国际化(I18N)可以使应用程序实现不同国家、不同地区语言环境的自动切换。

(5) 实现程序国际化有两种方式：跟随系统国际化和程序内国际化。

本 章 练 习

(1) Android 项目中，________和________文件夹用于存放不同的资源。

(2) res 目录中的资源在________类中被引用，并生成特定的 ID 值。

(3) 简述 Android 中常用的资源类型。

(4) 简述 Android 程序实现国际化功能的必要性。

附录　国家地区语言代码表

国家/地区	语言代码	国家/地区	语言代码
简体中文(中国)	zh-cn	繁体中文(台湾地区)	zh-tw
繁体中文(香港)	zh-hk	英语(香港)	en-hk
英语(美国)	en-us	英语(英国)	en-gb
英语(全球)	en-ww	英语(加拿大)	en-ca
英语(澳大利亚)	en-au	英语(爱尔兰)	en-ie
英语(芬兰)	en-fi	芬兰语(芬兰)	fi-fi
英语(丹麦)	en-dk	丹麦语(丹麦)	da-dk
英语(以色列)	en-il	希伯来语(以色列)	he-il
英语(南非)	en-za	英语(印度)	en-in
英语(挪威)	en-no	英语(新加坡)	en-sg
英语(新西兰)	en-nz	英语(印度尼西亚)	en-id
英语(菲律宾)	en-ph	英语(泰国)	en-th
英语(马来西亚)	en-my	英语(阿拉伯)	en-xa
韩文(韩国)	ko-kr	日语(日本)	ja-jp
荷兰语(荷兰)	nl-nl	荷兰语(比利时)	nl-be
葡萄牙语(葡萄牙)	pt-pt	葡萄牙语(巴西)	pt-br
法语(法国)	fr-fr	法语(卢森堡)	fr-lu
法语(瑞士)	fr-ch	法语(比利时)	fr-be
法语(加拿大)	fr-ca	西班牙语(拉丁美洲)	es-la
西班牙语(西班牙)	es-es	西班牙语(阿根廷)	es-ar
西班牙语(美国)	es-us	西班牙语(墨西哥)	es-mx
西班牙语(哥伦比亚)	es-co	西班牙语(波多黎各)	es-pr
德语(德国)	de-de	德语(奥地利)	de-at
德语(瑞士)	de-ch	俄语(俄罗斯)	ru-ru
意大利语(意大利)	it-it	希腊语(希腊)	el-gr
挪威语(挪威)	no-no	匈牙利语(匈牙利)	hu-hu
土耳其语(土耳其)	tr-tr	捷克语(捷克共和国)	cs-cz
斯洛文尼亚语	sl-sl	波兰语(波兰)	pl-pl
瑞典语(瑞典)	sv-se	西班牙语（智利)	es-cl